Sustainable Development Goals Series

The **Sustainable Development Goals Series** is Springer Nature's inaugural cross-imprint book series that addresses and supports the United Nations' seventeen Sustainable Development Goals. The series fosters comprehensive research focused on these global targets and endeavours to address some of society's greatest grand challenges. The SDGs are inherently multidisciplinary, and they bring people working across different fields together and working towards a common goal. In this spirit, the Sustainable Development Goals series is the first at Springer Nature to publish books under both the Springer and Palgrave Macmillan imprints, bringing the strengths of our imprints together.

The Sustainable Development Goals Series is organized into eighteen subseries: one subseries based around each of the seventeen respective Sustainable Development Goals, and an eighteenth subseries, "Connecting the Goals", which serves as a home for volumes addressing multiple goals or studying the SDGs as a whole. Each subseries is guided by an expert Subseries Advisor with years or decades of experience studying and addressing core components of their respective Goal.

The SDG Series has a remit as broad as the SDGs themselves, and contributions are welcome from scientists, academics, policymakers, and researchers working in fields related to any of the seventeen goals. If you are interested in contributing a monograph or curated volume to the series, please contact the Publishers: Zachary Romano [Springer; zachary.romano@springer.com] and Rachael Ballard [Palgrave Macmillan; rachael.ballard@palgrave.com].

Jang Singh

Sustainability and the Sustainable Development Goals in Cultural Clusters

A Comparative Analysis of Concepts, Influences, and Attainability

Springer

Jang Singh
Odette School of Business
University of Windsor
Windsor, ON, Canada

ISSN 2523-3084 ISSN 2523-3092 (electronic)
Sustainable Development Goals Series
ISBN 978-3-031-70617-2 ISBN 978-3-031-70618-9 (eBook)
https://doi.org/10.1007/978-3-031-70618-9

This Springer imprint is published by the registered company Springer Nature Switzerland AG
The registered company address is: Gewerbestrasse 11, 6330 Cham, Switzerland

Preface

News in the Summer of 2023 was dominated by reports of wildfires in Canada, Greece, the United States, and elsewhere. There were also storms of unprecedented magnitude. Governments around the globe were grappling with high inflation partly attributable to the Russia/Ukraine war. Migrants fleeing poverty and persecution in their homeland regularly lose their lives in the process. Environmental, economic, and social events like these have become common and underscore the need for renewed efforts to foster a sustainable future. This book builds on an earlier publication which was written when the world was in turmoil (Singh 2021): COVID-19 was wreaking havoc throughout the world, more so in some countries than in others. In many parts of the world, citizens were confined to their homes to help stop the spread of the virus. Families were separated and economies were shattered. However, amid the gloom, there were some bright, hopeful signs. For example, 1 week after Joe Biden was sworn in as President of the United States in 2021 among the first executive orders he issued were directives on racial justice and the United States rejoining the Paris Accord on Climate Change and a commitment to provide economic relief to Americans suffering hardship because of the pandemic. These three examples are directly linked to the three pillars of sustainability addressed in this book: the physical environment, social issues, and economic justice. President Biden clearly signaled that his administration was committed to a sustainable planet—a marked and welcome departure from the policies of his predecessor, Donald Trump. Before the advent of Trump and the pandemic, the world was making progress in achieving the Sustainability Development Goals established by the United Nations with the agreement of its member nations. However, it is unclear if the 17 goals will be met by the 2030 deadline. Those goals represented minimum targets for the global environmental, economic, and social pillars that underpin a sustainable future for our planet.

This book was written to gauge the views of a sample of political and business leaders from five countries on sustainability to determine if professional and national culture influenced their outlook. The author thought the views of political leaders were important because as policy makers their views could steer the direction of their constituencies. The views of business leaders warranted consideration because if they are at odds with those of the political leaders their views could jeopardize sustainability policies. Thus, this book reports and analyzes the views of 41 business and political leaders from five countries: the United States, Canada, India, Sri Lanka, and Guyana. The leaders were interviewed, and the transcripts of those interviews were analyzed to identify themes. These were then weaved into an assessment of progress being made on the UN Sustainability Development Goals (SDGs). In addition to primary data collected through the interviews, a review of the vast literature on sustainability-related issues allowed the author to present new data in consideration of progress being made in the quest for a sustainable future.

The conceptualization of this project began with clearance from the research ethics board of an academic institution. The process of conducting the interviews was enjoyable, challenging, and highly educational. It was not an easy task to arrange face-to-face interviews with leaders in five countries on three continents. However, the nature of the topic—sustainability—was of interest to the interviewees, and this paved the way for very interesting discussions. Weaving the themes uncovered with the SDG hopefully will add to our understanding of what it would take to achieve a sustainable future for our planet.

Windsor, ON, Canada Jang Singh

Reference

Singh J (2021) Sustainability: the influence of leadership and culture. Friesen Press, Altona.

Acknowledgments

The core of this book is interviews conducted with business and political leaders in five countries. These 21 political and 20 business leaders voluntarily granted interviews to the author largely because they care about the future of our planet. I am grateful to them for sharing their thoughts with me regarding the three pillars of sustainability: people, profit, and the planet. I am not at liberty to provide the names of the participants in the study because they were guaranteed anonymity and confidentiality as per the requirements of an academic research ethics board. I am also grateful to several individuals who helped me to arrange interviews. These individuals reside in the countries where the interviews were conducted. Again, I cannot name them, as doing so could lead to the identification of the interviewees. My research assistant for some of the literature on sustainability was Tara Chan, and I thank her.

My family has always been supportive of my endeavors and was very patient as I tried to explain concepts like qualitative research and the triple bottom line. I am grateful to my wife, Dr. Emily Carasco, whose decades of teaching, research, and activism helped to foster a generation more committed to sustainability than ours. Among that generation are my children: Elizabeth and David. To them, I dedicate this book.

Contents

Introduction

1

Abstract

This chapter describes the methodology employed in the study that constitutes the core of this book. It begins with an initial statement on the importance of sustainability and then describes the methods used to collect and analyze data on the study participants' views on the three pillars of sustainability. The rationale for selecting the participants from Guyana, Canada, India, Sri Lanka, and the United States of America is described and a summary of each of the 41 participants' views on the subject is provided.

1.1 Importance of Sustainability

Sustainability has emerged as a key concept in the business world, in the curriculum of business schools, and in society generally. However, what is clear is that the meaning of the term is not interpreted the same way by everyone. This book looks at the meaning of sustainability as interpreted by business and political leaders in five different cultures around the world.

What is the meaning of sustainability in extant literature? The origin of the term as it is used today is from the report of the United Nations World Commission on Environment and Development (WCED) published in 1987. The report titled *Our Common Future*, also known as the Brundtland Report, suggested that sustainable development is the kind of development that meets the needs of the present without compromising the ability of future generations to meet their own needs. The report noted that development is based on interdependence among nations as countries cannot develop in isolation. Therefore, the commission recommended that: "The pursuit of sustainable development requires a new orientation in international relations, and long-term sustainable growth will require far-reaching changes to produce trade, capital, and technology flows that are more equitable and better synchronized to environmental imperatives" (Brundtland 1987, p. 39). The environmental imperatives are necessary as "economic growth always brings risk of environmental damage, as it puts increased pressure on environmental resources" (Brundtland 1987, p. 39). Thus, the commission concluded that environmental protection is a key aspect of sustainable development, "as is a focus on the sources of environmental problems rather than the symptoms" (Brundtland 1987, p. 3). The commission linked the protection of the environment to economic justice as it noted that: "Those who are poor and hungry will often destroy their immediate environment in order to survive: They will cut down forests; their livestock will overgraze grasslands; they will overuse marginal land; and in growing numbers they will crowd into congested cities. The cumulative effect of these

J. Singh, *Sustainability and the Sustainable Development Goals in Cultural Clusters*, Sustainable Development Goals Series, https://doi.org/10.1007/978-3-031-70618-9_1

changes is so far-reaching as to make poverty itself a major global scourge" (Brundtland 1987, p. 29). This acknowledgment of intragenerational equity and the inherent intergenerational equity of a commitment to future generations provides the philosophical underpinning of the concept of sustainability as envisaged by the WCED.

1.2 How and Why Case Studies Were Selected

The study that forms the basis of this book examined the perceptions of sustainability of political and business leaders in five countries. These countries are Guyana, Canada, India, Sri Lanka, and the USA. Guyana was chosen to pilot the larger study, but the data gathered was significant enough to warrant dedicated analysis. In each of these countries, the intent was to interview five political and five business leaders for a total of 50 interviews overall. Comparisons were made between the views of political and business leaders and between these views in culturally different locations as defined by geographical location. Regarding the latter, the well-known cultural dimensions identified by Hofstede were used to identify culturally different jurisdictions.

However, culture is not only defined by geography but also by professional roles. In this instance, it is hypothesized that the role of political or business leader and geographical location would influence beliefs about sustainability. Various scholars have argued that using country as the sole determinant of culture may not fully explain peoples' beliefs, as these beliefs may also be influenced by factors such as the role they play in society. Carpar et al. (2015, p. 1012) suggest that there is "increasing evidence that country might be a less appropriate container of culture compared with other potential clustering dimensions such as age groups, socio-economic classes, or professional communities."

This study was focused on two such clusters: geographical regions—India/Sri Lanka, Canada/the USA, and role clusters—political leaders and business leaders. In particular, it was hypothesized that "individualism" would determine sustainability attitudes and practices, and therefore, the cluster of India and Sri Lanka would be similar, and the cluster of Canada and the USA would be similar, but the two clusters would be different. "The fundamental issue addressed by the 'individualism' dimension is *the degree of interdependence a society maintains among its members*. It has to do with whether people's self-image is defined in terms of 'I' or 'We'. In Individualist societies, people are supposed to look after themselves and their direct family only. In Collectivist societies, people belong to 'in groups' that take care of them in exchange for loyalty" (Geert--Hofstede.com).

It may also be argued that role influence attitudes and actions, and, therefore, the views and practices of business leaders will differ from those of political leaders. The study also investigated this proposition. Political leaders and leaders of institutions are thought to view the world differently. For example, Singh (1984) found a marked difference between the views of education leaders and political leaders in Guyana on the role of schools in national development. This difference is essentially a result of the ideological orientation of political leaders versus the pragmatic orientation of business and other nonpolitical leaders.

Other researchers have investigated the link between culture and sustainability. Miska et al. (2018) in one such study used the triple bottom line framework to test linkages between culture and the economic, environmental, and social pillars of sustainability. They found that the cultural dimensions of future orientation, gender equality, uncertainty avoidance, and power distance positively correlated with sustainability practices, while performance-oriented clusters negatively correlated with the pillars of sustainability. The findings of Miska et al. (2018) are not surprising. In a future-oriented culture, it is to be expected that attitudes toward sustainability would be positive, as sustainability is about the long-term survival of the planet. For those who are future-oriented, the value of sustainability practices would be evident. Similarly, the positive correlation found by Miska et al. between gender egalitarianism and sustainability is explained by the direct link between gender equality and the diminution of traditional gender roles in society.

Cultures that are characterized by high uncertainty avoidance would seek to avoid developments that threaten order in society. Therefore, these cultures would be positively disposed to sustainability measures. Power distance is another concept that is usually included in descriptions of culture and refers to the level of social stratification in a society. Miska et al. (2018) hypothesized and found support for the notion that high power distance will favor stronger sustainability practices. Their hypothesis is based on the assumption that in societies with well-defined power structures and where those in authority are inclined to promote sustainability such practices would be successful. However, one might also argue that if those with authority in high power distance societies are not inclined to promote sustainability, the reverse would be true. Moreover, it could also be argued that in such cultures there would be a strong tendency to maintain the status quo at the expense of social, economic, and environmental sustainability.

1.2.1 National Culture

Hofstede defines culture as "the collective programming of the mind distinguishing the members of one group or category of people from others" (Hofstede Insights 2020). Geert Hofstede who conducted a large global study on the link between culture and organizational values initially identified four dimensions of national culture. This initial study was conducted between 1967 and 1973 in 70 countries with International Business Machines Corporation (IBM) offices. Subsequent research with others led to the addition of two dimensions. Hofstede labelled the original four dimensions of culture as: Power Distance Index (PDI), Individualism Versus Collectivism (IDV), Masculinity Versus Femininity (MAS), and Uncertainty Avoidance Index (UAI). The two additional dimensions were labelled: Long-Term Orientation Versus Short-Term Normative Orientation (LTO) and Indulgence Versus Restraint (IVR). These dimensions were identified through empirical research with respondents completing questionnaires and the collected data being processed statistically. The resulting scores on the various dimensions for different countries indicate the standing of countries relative to each other.

Power Distance Index (PDI) is a measure of the social stratification in a particular society. "People in societies exhibiting a large degree of Power Distance accept a hierarchical order in which everybody has a place and which needs no further justification. In societies with low Power Distance, people strive to equalize the distribution of power and demand justification for inequalities of power" (Hofstede Insights 2020). The clear expectation here is that in high PDI countries, there would be resistance to measures aimed at social and economic sustainability.

Individualism refers to a concern primarily with one's interests as opposed to those of groups extending beyond his/her immediate family. In highly individualistic countries like the USA and Canada, this cultural dimension suggests that generally people are focused primarily on their self-interest and those of just immediate family members. In such societies, it could be expected that there would be resistance to initiatives aimed at all three pillars of sustainability.

Masculinity Versus Femininity (MAS) assigns traditional gender roles to particular cultures. Cultures high in masculinity would be more competitive with emphasis being placed on material rewards for success, while cultures low on masculinity would be consensus-oriented and emphasize cooperation rather than competition. Femininity is concerned with quality of life over materialism. One may surmise that a more feminine culture would be amenable to social, economic, and environmental sustainability, reflecting a concern for quality of life over materialism.

Uncertainty Avoidance Index (UAI) measures the risk level of cultures. Those cultures high on the Uncertainty Avoidance Index would avoid uncertainty and ambiguity. In such cultures, there would be adherence to principles aimed at controlling the future and not engaging in nontraditional behavior. This would suggest that cultures high on uncertainty avoidance would be more conservative and likely to favor environmental protection but be resistant to social and economic sustainability.

The two dimensions of culture added to Hofstede's original four are Long-Term Orientation (LTO) and Indulgence. All cultures must be concerned with the future while addressing issues of the present. Different cultures address these issues in different ways. Those who score low on this dimension would be more wedded to tradition and resistant to change in dealing with the future. However, those who are more long-term in their thinking would embrace the changes necessary to survive and thrive in the future. It would be reasonable to expect that cultures scoring low on Long-Term Orientation would be resistant to all the pillars of sustainability, while those with high scores on this dimension would embrace sustainability initiatives.

Indulgence is a measure "of basic and natural human drives related to enjoying life and having fun. Restraint stands for a society that suppresses gratification of needs and regulates it by means of strict social norms" (Hofstede Insights 2020).

Hofstede Insights (2020) provides data on the various dimensions of culture for many countries in a database developed through numerous empirical studies. The database includes information on four of the five countries included in this study. It includes data on Canada, the USA, India, and Sri Lanka, but it does not include information on Guyana. However, Guyana is culturally similar to Trinidad and Tobago, and the data from that country is included to show how Guyana might score on the dimensions of culture.

1.2.2 Dimensions of Culture: Trinidad and Tobago

The following are Trinidad and Tobago's scores for the dimensions of cultures according to Hofstede Insights (2020) (Fig. 1.1):

- Power Distance: Trinidad and Tobago scores 47 on Power Distance, which indicates, among other things, equal rights, superiors are accessible, power is decentralized, and control is disliked.
- Individualism: Trinidad and Tobago is a collectivistic society, as it has a low score of 16 on Individualism. In collectivistic societies, one could expect to find long-term commitment to groups with a premium placed on loyalty and looking after the welfare of group members.
- Masculinity: Trinidad and Tobago scores 58 on Masculinity and is therefore regarded as a Masculine society. This implies that it is a society driven by competition and where success is indicated by winning. It is not a feminine society, where success is defined by quality of life and caring for others.
- Uncertainty Avoidance: Trinidad's intermediate score of 55 on Uncertainty Avoidance suggests a middle-of-the-road approach between trying to control the future and leaving things to chance.
- Long-Term Orientation: Trinidad and Tobago has a low score of 13 on this dimension. It is therefore classified as a normative culture, meaning that it values traditions and is resistant to societal change. In normative societies, it should be expected that a premium is placed on quick results versus on saving for the future.
- Indulgence: Trinidad and Tobago has a very high score of 80 on Indulgence. As Hofstede (2020) points out, people in such societies "possess a positive attitude and have a tendency towards optimism. In addition, they place a higher degree of importance on leisure time, act as they please, and spend money as they wish."

1.2.3 Dimensions of Culture for Canada, India, Sri Lanka, and the USA

The scores from Hofstede Insights (2020) on the different dimensions of culture for the countries covered in this study are shown in Fig. 1.2. On all the dimensions, the similarity of the scores for Canada and the USA should be noted, as should

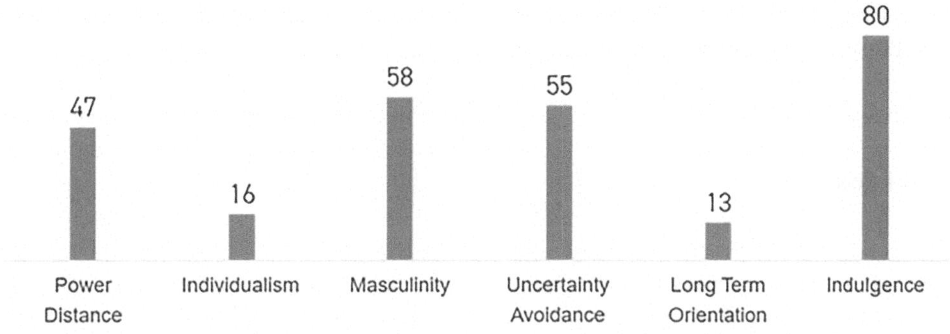

Fig. 1.1 Trinidad and Tobago

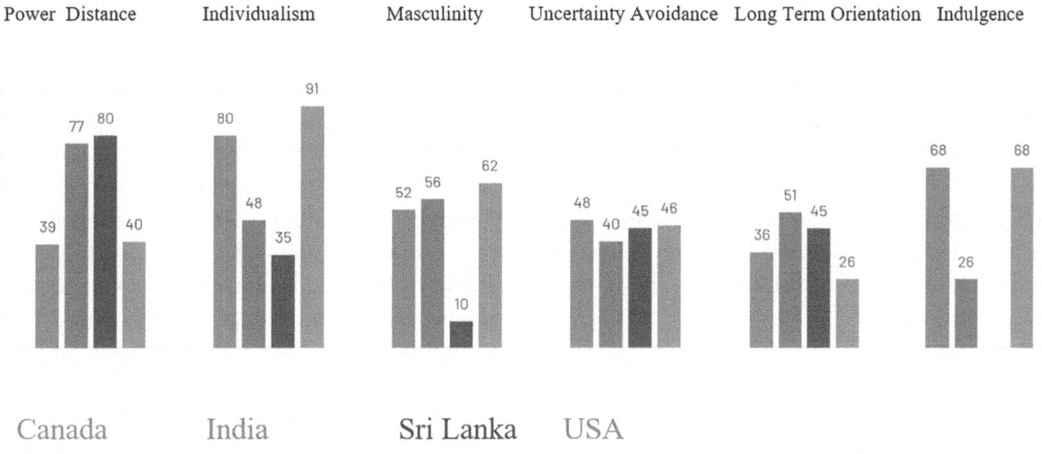

Fig. 1.2 Canada, India, Sri Lanka, and the USA

the scores for India and Sri Lanka. It is also striking how different the North American scores are from the Asian scores.

1.2.3.1 Canada

Canada's score of 39 on Power Distance indicates that Canadian culture values egalitarianism. The score of 80 on Individualism suggests a highly individualistic society, while its score of 52 on Masculinity points to a moderately competitive society in which winning at all costs is not highly valued. Canada's score of 48 on Uncertainty Avoidance suggests a culture that is not threatened by innovation and is open to new ideas in dealing with the future. However, the country's score of 36 on Long-Term Orientation indicates that while not threatened by innovation, Canadians are generally not long-term oriented. This suggests a propensity for quick results and a low savings rate. The overall score of 68 on Indulgence is indicative of an impulsive and optimistic people.

1.2.3.2 India

India, as may be expected, scores high on Power Distance. Its score of 77 points to a highly stratified society where power is centralized, and communication is top-down. At the same time, its Individualism score of 48 suggests a society with both individualistic and collectivistic properties. A Masculinity score of 56 puts India in the highly masculine category where success is measured by material rewards. An Uncertainty Avoidance score of 40 puts the country in the medium to low category for avoiding uncertainty. This means that the culture tolerates the unexpected. India scores 51 on Long-Term Orientation, which suggests a middle-of-the-road view of the future. On Indulgence, the country is classified as one of restraint with a score of 26.

1.2.3.3 Sri Lanka

Like India, Sri Lanka with a Power Distance score of 80 is a hierarchical culture that tolerates inequalities. However, it is simultaneously a collectivistic society with a low score of 35 on Individualism, suggesting loyalty is highly valued. Sri Lanka has a very low score of 10 on Masculinity, making it a Feminine society as per Hofstede's classification. On Uncertainty Avoidance, the country is middle of the road with a score of 45. This signals no preference for either a desire to control the future or to just let it unfold. Similarly, the country has an intermediate score of 45 on Long-Term Orientation. Thus, traditions are valued but so is preparation for the future. The Hofstede database does not yet have a score for Indulgence for Sri Lanka.

1.2.3.4 USA

The USA has a score of 40 on Power Distance and a score of 91 on Individualism, suggesting that it is a society where superiors are accessible, but the emphasis is on the individual over groups. It also has a high score of 62 on Masculinity, meaning that success is measured in material terms. A low score of 46 on Uncertainty Avoidance suggests that the USA is not focused on controlling the future but is open to letting it just unfold. The country is also not long-term in its orientation, scoring just 26 on this dimension of culture. This perhaps explains the emphasis in US business on quarterly results. Like Canada and unlike India, the USA with a score of 68 is classified as an indulgent society where there is only moderate restraint on desires and impulses.

1.3 Data Collection and Analysis

Sustainability as defined by John Elkington (1997) has three pillars. Much of the current discussion of this topic revolves around the Elkington framework of the economic, environmental, and social pillars. As illustrated previously, global sustainability is in serious peril, as there are serious problems in each pillar of sustainability. The literature and statistics on these problems are manifold. Some of the major sustainability challenges will be addressed in the relevant sections on sustainability in the five countries discussed in this book. The key question that this study sought to answer was: How do the business and political leaders of Guyana, Canada, India, Sri Lanka, and the USA perceive the concept of sustainability? How do they define the concept? And how do they view particular aspects of each pillar of sustainability in their respective countries?

Answers to these questions were sought through face-to-face interviews that were recorded and later transcribed and analyzed. The focus was on identifying and expounding identified themes. Such studies are classified as qualitative research. Like all scientific research, qualitative studies systematically seek to answer questions by collecting data, revealing new findings, and including elements of generalizability. "Qualitative research is multimethod in focus, involving an interpretative, naturalistic approach to its subject matter. This means that qualitative researchers study things in their natural settings, attempting to make sense of, or interpret, phenomena in terms of the meanings people bring to them. Qualitative research involves the studied use and collection of a variety of empirical materials—case study, personal experience, introspective, life story, interview, observational, historical, interactional, and visual texts—that describe routine and problematic moments and meanings in individuals' lives" (Denzin and Lincoln 2005, p. 2).

Qualitative research differs from quantitative research in what it seeks to do and the methodologies used. These differences are summarized by Mack et al. (2005) as shown in Table 1.1. The key difference between quantitative and qualitative research is that qualitative research allows for flexibility in the questions asked and for in-depth questioning on issues of interest. Quantitative methods are generally inflexible. They usually begin with hypotheses that are

Table 1.1 Comparison of quantitative and qualitative research approaches

	Quantitative	Qualitative
General framework	Seek to confirm hypotheses about	Seek to explore phenomena
	Instruments use more rigid style eliciting and categorizing responses to questions	Instruments use more flexible, of iterative style of eliciting and categorizing responses to questions
	Use highly structured methods as questionnaires, surveys, and structured observation	Use semi-structured methods such as in-depth interviews, focus groups, and participant observation
Analytical objectives	To quantify variation	To describe variation
	To predict causal relationships	To describe and explain relationships
	To describe characteristics of a population	To describe individual experiences
		To describe group norms
Question format	Closed-ended	Open-ended
Data format	Numerical (obtained by assigning numerical values to responses)	Textual (obtained from audiotapes, videotapes, and field notes)
Flexibility in study design	Study design is stable from beginning to end	Some aspects of the study are flexible (for example, the addition, exclusion, or wording of particular interview questions)
	Participant responses do not influence or determine how and questions researchers ask next	Participant responses affect how which questions researchers which ask next
	Study design is subject to statistical assumptions and conditions	Study design is iterative, that is, data collection and research questions are adjusted according to what is learned

Mack et al. (2005, p. 3)

tested in selected, representative samples with findings generalized to a population. By their design, they are not meant to pursue leads uncovered during a study, hence the inflexibility label. The qualitative approach is better suited for studies such as exploring the views of business and political leaders on sustainability.

While much has been written on sustainability—how it may be measured and what needs to be done to achieve its objectives—to the best of the author's knowledge, no one has systematically explored the views of business and political leaders on basic questions relating to sustainability. It is important to examine those views, as these are the people who are key in implementing the grand objectives of initiatives like the United Nations' Sustainability Development Goals. Do they even know what these goals are? Are they familiar with John Elkington's triple bottom line? In the absence of hard data on questions like this, it would be difficult to formulate meaningful hypotheses to be tested in a formal, quantitative way using statistical techniques. Moreover, the richness of the data collected through face-to-face interviews as in this study could provide very valuable information on practical steps to be taken in pursuing sustainability goals. Qualitative studies such as this provide richer data with a deeper understanding of the issue being investigated. Moreover, they provide more respect for respondents, partly by allowing the researcher to further explore unanticipated answers. Thus, the respondent and researcher are not confined to predetermined questions and answers but are free to probe interesting perspectives.

1.3.1 Study Participants

To ascertain the views of the two categories of leaders in the various countries, it was decided that five respondents from each category for each country would be interviewed. These numbers were thought sufficient to gain a thorough understanding of the perspectives of the respondents according to culture and role. The interview as the medium to elicit these perspectives is critical to success in studies of this nature. The crucial advantage of the interview is that it allows the researcher to pursue leads that may arise during the interaction. Asking direct questions is a "strength because a great deal of information needed in social scientific research can be gotten from respondents by direct questions" and if these questions are correctly handled, even personal or controversial material can be successfully obtained (Kerlinger 1964, p. 479).

As previously noted, the population for the study—the sample universe—was political and business leaders in Guyana, Canada, Sri Lanka, India, and the USA. The former includes elected members (past or present) of the national assembly, heads of municipalities, and candidates for national or state office. The latter includes officers (past or present) of major corporations and heads of business federations. These parameters guided the selection of prospective interviewees.

A purposive sampling strategy was used. Robinson (2014, p. 32) describes this approach as a nonrandom way of "ensuring that particular categories of cases within a sampling universe are represented in the final sample of the project." In the present study, a target of five leaders from each of the two categories was set, but because of logistical reasons, the target was not always achieved. However, a total of 41 interviews of 21 political leaders and 20 business leaders was successfully completed. The breakdown of the interviewees by role and country is shown in Table 1.2.

The respondents were recruited mainly through personal contacts, i.e., the researcher used his network to arrange the interviews through intermediaries. Although direct approaches were initially made, they yielded few positive responses.

Table 1.2 Sample by country and role

	Political	Business	Total
Guyana	5	4	9
Canada	3	5	8
India	6	4	10
Sri Lanka	3	3	6
USA	4	4	8
Total	21	20	41

The data analysis procedures for this study involved a thematic narrative analysis. The analyses followed the outline of procedures by Riessman (2008). A thematic narrative analysis aims to accomplish an integrated and multidimensional explanation of human experiences. The dimension of time differentiates thematic narrative analysis from thematic analysis and a grounded theory. Temporal elements preserved the sequencing of events in the participants' experiences.

The sample of this study was grouped by the country where the leaders were from, and then further grouped into political and business leaders, yielding 10 groups in the sample. The groups were used to conduct an intercultural analysis, an inter-role analysis, and an intercultural and inter-role analysis to compare the perceptions of the participants on sustainability. For the purposes of this study, sustainability was examined on three pillars, which were: economic, environmental, and social (Elkington 1997).

1.3.2 Data Analysis Methods

According to Riessman (2008), thematic narrative analysis emphasizes what the participants say rather than how, why, or to whom the participants direct what they say. Therefore, the first step of the analysis was to read and re-read the transcripts and identify chunks of text that meaningfully represent experiences with sustainability. Nonetheless, the analysis also emphasizes the preservation of the "whole story" through integrating chronology (Riessman 2008). Thus, the second step of analysis involved categorizing the chunks of text into abstract themes associated with the temporal elements of sustainability experiences. To conduct the first two steps of analysis, the transcripts were uploaded to NVivo 12 Pro, a qualitative data analysis software. The software allowed the researcher to highlight chunks of data and assign them to nodes, a feature of the software often used to represent the smallest unit of meaning. For instance, GB1 (Guyana Business Leader 1) narrated:

> *Immediately what comes to mind is the long-term profitability, taking the long view and tying it in with ethics. It obviously means, I would think, having a very conservative view of your strategies, not looking for the quick buck, as we say. But looking to position your business so that over the long-term, you can continue to add value, build value, for your stakeholders.*

In this chunk of data, the participant referred to sustainability in relation to making a profit; hence, the chunk of data was highlighted and assigned to the node "making a profit." However, to reflect the temporal element, GB1's narrative told the story of not just making a profit but making a profit over some time. As a result, the initial theme "long-term profitability" emerged, as making a profit over time was considered to be part of GB1's experience with sustainability. Initial themes for business leaders and political leaders were generated separately to allow comparison between the experiences of both groups of participants.

In the third step of the analysis, the initial themes were then reviewed against the raw data to ensure that the themes were consistent and explicitly represented in the data. The initial themes for business leaders and political leaders were integrated to develop the final themes.

1.3.2.1 Intercultural Analysis

The sample involved political and business leaders from Guyana, Canada, India, Sri Lanka, and the USA. Both political and business leaders from each country were treated as a unit to be able to cross-examine the themes from the five countries. Perceptions of the influence of the local culture, tradition, history, and religion on sustainability were among the items compared in the intercultural analysis.

1.3.2.2 Inter-role Analysis

In the inter-role analysis, the participants were assigned into two groups: political leaders and business leaders. Regardless of the leaders' country of origin, the leaders' perceptions as political or business leaders were examined as a whole. The analysis involved the examination of how the context in which leaders worked influenced their perceptions of sustainability.

1.3.2.3 Intercultural and Inter-role Analysis

The perceptions of each group of leaders from the five countries were then compared to examine similarities and differences in perceptions of sustainability. The analysis also involved examining the influence of culture on leadership roles in terms of sustainability. The analysis was conducted in the form of a matrix.

1.3.3 Results by Participant

This section contains the results from the narrative analysis. Findings from the narratives of each participant are provided in the subsections below. With some participants being high-profile politicians, some information such as current and previous positions have been deliberately omitted to protect the participants' identities. The findings were focused on the temporality and sequencing of the narratives—that is, what the participants were saying in terms of their experiences, rather than on thematic meanings (Riessman 2008).

The key used to identify participants is as follows: GB (Guyana Business), GP (Guyana Political), CB (Canada Business), CP (Canada Political), IB (India Business), IP (India Political), SB (Sri Lanka Business), SP (Sri Lanka Political), UB (United States Business), UP (United States Political).

GB1 GB1 has been a banker for 40 years. The participant was focused on long-term profitability rather than quick returns. However, bankers also need to take investors into account. In terms of social sustainability, the participant narrated that ethics came into play in deciding whether to have a budget to satisfy current needs (i.e., raising salaries, opening more jobs) or to set money aside for future needs. For planet sustainability, the participant perceived that Guyana's culture contributes to the way locals perceive natural resources. Locals see natural resources as opportunities for profit and are not environmentally conscious. Explaining sustainability to locals might prove to be a difficult and expensive task; nonetheless, the government is leading a movement towards environmental consciousness.

GB2 GB2 articulated his perception of the definition of sustainability as the ability of a business to maintain profitability over a period of time. Long-term planning was considered essential to sustainability. The government did not lack legislation and policies for sustainability. As a businessman, the participant perceived that such legislation and policies might be restrictive to his income. Nonetheless, the participant continued to comply with the rules, as he believed responsibility and trust were the foundation of a stable business. In terms of social sustainability, the participant perceived that the free primary and secondary education and the inexpensive tertiary education in Guyana provided employment opportunities for the locals. However, the participant believed that the workforce needed to be more diverse in terms of gender and race. For planet sustainability, the participant shared that the government and private sector were beginning to take steps to be environmentally conscious.

GB3 GB3 is the president of an organization representing the private sector. The participant perceived that sustainability in a business involved a process leading to a business being able to run itself and continuously generate profit. The participant spoke about the private sector being able to balance generating returns on investments while allowing people to maintain a reasonable standard of living and leaving the environment habitable for future generations. The participant believed that either education or an existing family business provided opportunities for generating income and employment. Environmentally, the participant reiterated that natural resources must be taken care of as one takes care of his assets.

GB4 GB4 retired as chairman but is currently a board member of a distillery company. The participant narrated his extensive experiences with the sugar industry before entering the distillery business. The participant recalled that long-term planning helped him open and sustain his distillery. Even with the success of his company, the participant revealed that he was continuously planning for the future. The participant also shared that he offered shares to the public for income opportunities.

GP1 GP1 is a former Prime Minister of Guyana and is highly focused on the political aspect of social sustainability. The participant perceived that whatever happened to the developed countries in Europe and North America might not apply to Guyana. The participant believed that sustainability is a long-term process, and one can only learn from one's own experiences. According to the participant, people need to be able to adapt to changes to be able to be sustainable.

GP2 GP2 is the head of a major government commission with strong links to the environment. GP2 perceived sustainability as a responsibility of the government, businesspersons, and the public. The participant believed that the general public plays a role in sustainability, specifically in planet sustainability. The government was responsible for making policies to protect the public's interests. The participant explained that in the field of forestry, the industry was fixed in a single location; therefore, the government urged businesses in the forestry industry to contribute to the local communities. Overall, the participant shared the government's initiatives to promote corporate social responsibility.

GP3 GP3 is a former Prime Minister of Guyana. GP3 perceived that sustainability in general was the nation's continuous ability to achieve goals. For the participant, planning and budgeting were important parts of development, and specifically, realistic long-term planning that will benefit the people. The participant shared that the history of

Guyana as a colony has affected the country's economic development. Despite possessing land and natural resources, the country has a colonial mindset and has not allocated its resources properly. While the government has been taking action to support the locals, the participant stated that such expenses might not be sustainable in the long run.

GP4 GP4 is a former speaker of the house and an executive member of a political party. The participant believed that the government was responsible for educating people about sustainable methods. While the participant admitted to a lack of knowledge of the formal definition of economic, social, and planet sustainability, the participant also stated that he understood how profitability and sustainability operated, starting with investing in education. Politically, the participant believed that unequal distribution of wealth in any country could lead to an imbalance in power, citing instances in the USA, the United Kingdom, and France. The participant also stated that political instability can be the result of power leaning extremely to one side. The participant finally claimed that Guyana has learned from those countries and has taken action to achieve a "green economy."

GP5 GP5 is a former minister of finance in Guyana. The participant mentioned that Guyana recently experienced a change of government, and the change might have affected profitability. The inability of businesses to cope with the change was perceived to be due to the lack of sustainability. Socially, the participant perceived that income inequity and employment inequity were becoming less of an issue in Guyana, as the participant believed that the gap between the rich and poor was closing. Environmentally, the participant also perceived that issues with pollution or climate change did not exist following the agreement with the Norwegian government, in which Guyana was to receive rewards for cutting down pollution.

CB1 CB1 is the CFO of a major Canadian corporation in the entertainment field. As a businessman in Canada, CB1 believed that sustainability involves the ability to "pass the corporation on to the next generation." The participant believed in human capital. As such, CB1 believed that Canadian workers were not experiencing income inequality as much as workers from other countries. In terms of employment equity, the participant reiterated that as a father to two daughters, he appreciated the "diversity of opinions" on the board. The participant also emphasized that diversity was not limited to gender, but also to race, religion, and age. For environmental sustainability, CB1 believed that the implementation of government legislation was important so that businesses had guidelines.

CB2 CB2 was the CEO of several of Canada's most successful corporations. For CB2, sustainability involves being able to finance sustainable materials, resources, and infrastructure. The participant also emphasized that having a large percentage of people in poverty was not economically sustainable and provided the situation of the USA as an example. In terms of social sustainability, CB2 perceived that Toronto, in particular, was "inclusive." For environmental sustainability, the participant mentioned finishing the Al Gore course for Climate Reality Change, and that he has been working towards fighting against climate change.

CB3 CB3 is the CEO of a mid-size Canadian corporation in the construction industry. The participant believed that the three pillars of sustainability were interrelated. CB3 explained that to be able to maximize economic activities, people and natural resources needed to be taken care of. However, the participant believed that some form of income and employment inequality still existed. The participant suggested that transparency would deter inequalities. CB3 thought that organizational culture played an important role in avoiding inequalities. Environmentally, the participant acknowledged climate change and global warming. While CB3 believed that businesses could take action to save the environment, the participant emphasized the impact of individual lifestyles, such as waste management, buying sustainable products, and commuting to work.

CB4 CB4 believed that economic sustainability involved innovation. Adapting to changes was considered sustainable. Socially, the participant believed that women experienced income inequity. The participant argued that in his business, people get paid according to their position, not their gender or race. Environmentally, CB4 believed that everyone was responsible for taking care of natural resources and referred to the environment as a "social constant."

CB5 CB5 was the head of a business association. CB5 claimed that economic sustainability was most relatable to her as a person in business. To be economically sustainable, the participant believed that the business should be able to adapt to changes. The participant also believed in human resources not only for economic sustainability but also for social sustainability. CB5 mentioned that excluding certain genders or races meant that not all human resources were being utilized, which is not ideal for economic growth. For environmental sustainability, CB5 stated that there needs to be a balance between making a profit and preserving natural resources.

CP1 CP1 was the mayor of a mid-size city at the time of the study. The participant believed that a huge wage gap was not economically and politically sustainable, as people would most likely be discontented. Nonetheless, the participant recognized that the top 1% influenced the local and federal government. However, as a government leader, the participant trusted the rich as business leaders who wanted to be the voice of their employees and the community. Socially, CP1 did not believe that women experienced employment inequity and cited women with high positions in the government. For environmental sustainability, CP1 perceived that urban planning was important.

CP2 CP2 was the mayor of a small municipality at the time of the study. The participant discussed residential development as the majority of the tax base and associated the taxes with economic and social sustainability. As the residential areas developed, the participant believed that the area became more economically viable. Socially, the demographics of the residents were largely heterogeneous. Environmentally, the participant believed in investing in sustainable equipment, such as solar panels.

CP3 CP3 is one of the longest-serving members of Canada's federal parliament. As a political leader, CP3 believed that economic sustainability involved being competitive in the global market. The participant also believed in closing the wage gap to be able in order to be economically sustainable. The participant claimed that Canada as a developed country had an advantage over developing countries in terms of providing equal employment and income opportunities. However, CP3 believed that as a developed country, Canada was a "role model" and needed to support developing countries. CP3 stated that humans as a species shared a collective responsibility towards each other and the environment.

IB1 IB1 is one of the most prominent businesspersons in a small Indian state. IB1 believed that India was a poor country; however, India's economy has been growing for the past 15 years. IB1 revealed that India was still experiencing issues with education, economics, and utilizing resources; nonetheless, the situation was considered to be better than it was in the past. IB1 believed that the "imbalance" in the social aspect was hindering the growth of India. A small percentage of the population generated income taxes. In terms of income inequity, the participant claimed that people from certain areas received less wages. The caste system was not much of an issue in IB1's local community due to the colonial influences of the Portuguese. The participant discussed the divide between North and South India and believed that Northern India experienced more inequalities and was less environmentally sustainable.

IB2 IB2 was the CEO of a major public-sector corporation. IB2 claimed that India was not yet economically sustainable; however, the economy has been continuing to grow in the past few years. IB2 mentioned the 1991 economic reforms and believed the reforms to be the source of growth. Businesses were utilizing men and women in the workforce; however, the caste system was still a source of discrimination. IB2 also emphasized that the government was writing legislation to solve problems with pollution and illegal logging.

IB3 IB3 is the youthful CEO of a successful renewable energy corporation. IB3 emphasized his role as a businessman and that the viability of the business was his priority. Nonetheless, the participant believed that viable business models included proposals for economic, social, and environmental sustainability. When discussing income inequity, IB3 mentioned the global competition in which developed countries tend to pay higher wages than developing countries. As society becomes more globalized, the participant believed that income inequality would increase. IB3 also believed that the continuous development of technology contributed to environmental sustainability and provided the example of the wide use of solar panels in India and China.

IB4 IB4 is a successful business consultant and business owner. IB4 explained that taking care of customers and the local community contributed to both social and economic sustainability. As a person in business, IB4 believed that religion did not contribute to providing opportunities. The participant emphasized the competence of an individual to be more vital than religion. Quality and technology were two factors that the participant believed contributed to environmental sustainability. Given the nature of the participant's business, the customers were knowledgeable on how to utilize resources well.

IP1 IP1 believed that India has several natural resources and products for export. However, expenses have been largely related to fuel. The participant also mentioned that situations were different in different parts of the country. The participant cited the western part of India as suffering from a lack of water supply. A large portion of the population was living in poverty, and the participant believed that the number was as large as the entire US population.

IP2 IP2 mentioned agriculture as a main source of livelihood for a large portion of the population. The lack of government policies to regulate the proper utilization of farmlands was believed to be the source of destruction. IP2 argued that the implementation of environmental laws should be stricter for increased sustainability.

IP3 IP3 was an elected official at both the State and Central levels. IP3 believed that inequality existed not just in India. However, the participant believed that everything was created to be unequal—"like all five fingers are not equal." The participant also believed that nature had power on its own and that nature controlled human actions more than human actions controlled nature. Nonetheless, the development of infrastructure and roadwork was necessary to attract tourists and grow the economy. IP3 believed that finding a balance was key to sustainability.

IP4 As a political leader, IP4 believed that the government was attempting to have its programs reach all Indians. The government has also been making sustainable policies for the past 70 years. Sustainable policies included economic and social aspects.

IP5 IP5 believed that the development of roads might increase economic and environmental sustainability. In developing roads, the participant believed that transportation costs would lessen, and the emission of carbon would also lessen. As a developing country, the participant believed that opportunities for income and employment have been rising. The government has initiatives to help families in terms of health so that the working class can focus on working.

IP6 Recent construction in the business sector was believed to be environmentally friendly. IP6 perceived that sustainability should play a role in decision-making, not just in business, but in everyday lives. Notably, IP6 related Indian principles from the Vedas to the current concept of sustainability.

SB1 SB1 is a successful business consultant and business owner. SB1 claimed that Sri Lankans were not initially materialistic. Colonial influences contributed to the materialistic wants of the locals. Consumerism has become prevalent. The participant perceived consumerism to be somewhat unsustainable, as resources were being exhausted. The participant believed that people need to be re-educated about living simple lives.

SB2 SB2 was a successful businessperson and former head of a large public utility. For SB2, making a profit constituted economic sustainability. The participant emphasized the ability to adapt to changes as a contributing factor to a sustainable business. As an example, the participant cited the dependence on oil, and how oil prices were fluctuating. Finding alternative energy might be a solution to changing oil prices. The status of the business was also believed to rely on globalization and future generations. Socially, SB2 believed that Sri Lankans did not experience as much inequality as Indians. The participant then acknowledged the existence of climate change and how climate change negatively affected agriculture.

SB3 SB3 is one of Sri Lanka's most successful businesspersons with interests in some of the country's largest corporations. As a businessperson, SB3 believed that sustainability was related to finances. SB3 perceived that corruption was widespread in Sri Lanka. The participant expressed that corruption impacted the success of a business. In terms of the middle class, SB3 believed that the increase in purchasing power was economically and socially sustainable. Lastly, the participant believed that the work of previous government regimes increased the protection of the environment and considered the work to be a "good investment."

SP1 SP1 is a former minister and member of Sri Lanka's Parliament. SP1 perceived that the gap between the richest and the poorest people was wide. However, the gap was more visible in urban areas than in rural areas. SP1 stated that rural areas were more socially sustainable, as people had sources of income, and they grew their own food. Nonetheless, members of the younger generation were not satisfied and chose to move to urban areas, which the participant perceived as not sustainable. Due to the lack of jobs, people turned to illegal drugs. Environmentally, SP1 related Buddhist principles to sustainability. SP1 stated that according to Buddhist principles, everything in the environment was interrelated.

SP2 SP2 was a member of Sri Lanka's parliament. SP2 believed that the previous generation experienced an industrial revolution, while the current generation was experiencing a technological revolution. To be sustainable, people needed to adapt to technology. The participant believed that Sri Lanka, as a developing country, was experiencing an unsustainable market. SP2 perceived that developed countries "owned a lot" and might be monopolizing the market. Developing countries generally provide cheap labor.

SP3 SP3 is a social/environmental activist with a national profile. SP3 mentioned that organic farming was currently profitable, which was sustainable economically and environmentally. The participant also believed that whatever was profitable was socially sustainable, as the profits benefit the people. SP3 related sustainability to the Buddhist lifestyle, in which everything in the environment was connected, and humans were not superior but equal to the environment.

UB1 UBI was the head of a major Business Association. UB1 was of the view that using sustainable materials had positive impacts on the environment and business. Using sustainable materials lessened operational expenses. UB1 also thought that the younger generation seemed to be more environmentally conscious and would not support businesses reported to be harming the environment.

UB2 UB2 is an Executive in the Health Care Industry. UB2 argued that sustainability was a global matter. Therefore, the participant believed that income inequality was dominant in some parts of the world, such as India. As a developed country, the USA had to adjust to help developing countries.

UB3 UB3 is the CEO of many successful businesses in California. UB3 believed that environmental policies governing businesses were stricter than they had been. Abiding by the law was perceived to be an economically sustainable action, as penalties were expensive. The participant's business made changes to adapt to the policies. In terms of social sustainability, UB3 believed that employment inequity has been significantly reduced. Women were frontrunners in a major part of his businesses.

UB4 UB4 is a senior Executive in the Health Care field. UB4 reiterated that basic education was important for sustainability. Educated citizens were more likely to secure jobs and earn "economic freedom." Concerning education, the participant stated that more women were being educated. As a result, more women entered the workforce, minimizing the employment inequity in terms of gender. UB4 believed that every person has a social and environmental responsibility and that individual actions contribute to social and environmental growth.

UP1 UP1, a mayor, perceived that education was essential to economic and social sustainability. As a political leader, the participant believed that the government was responsible for taking care of the people and that providing education was a way to secure people's futures. UP1 also believed that God provided people with natural resources; however, taking care of the resources was the people's responsibility. The government was taking action to protect the environment through legislation.

UP2 UP2 was the mayor of a medium-sized city in New England. UP2 believed that the existing wealth gap was related to the education gap. The rate of unemployed people was high. Therefore, the participant claimed that providing education might help them to become self-sufficient, which was sustainable economically and socially. As a political leader, the participant was also striving to create more jobs. UP2 also believed that environmentally sustainable materials, such as solar panels, helped conserve energy and reduce expenses.

UP3 UP3 is a State Senator. UP3 claimed that a wealth gap existed all over the world and that education might be serve to reduce the gap. Education was believed to open employment opportunities. The participant was working on providing equal access to education.

UP4 UP4 is a politically appointed county official. UP4 believed that resources were limited and that taking care of the resources was the responsibility of the current generation to sustain future generations. The participant was using his power as a political leader to develop plans and policies to conserve resources. The participant was working on a waste management policy, focusing on recycling and renewable energy.

This chapter discussed the importance of sustainability and described the methods used to collect and analyze data on the participants' views on the three pillars of sustainability. It also provided a summary of each of the 41 participants' views on the subject. In the next nine chapters, these views will be examined in greater detail with relevant background information for the countries where interviews were conducted: Guyana, Canada, India, Sri Lanka, and the United States of America.

References

Bruntland Report (1987) Our common future: Report of the World Commission for Environment and Development. WCED, New York

Carpar DV, Devinney TM, Kirkman BI, Caligiuri P (2015) Conceptualizing and measuring culture in interna-

tional business and management: From challenges to potential solutions. J Int Bus Stud 46(9):1011–1027
Denzin NK, Lincoln YS (eds) (2005) The Sage handbook of qualitative research, 3rd edn. Sage Publications Ltd., Thousand Oaks
Elkington J (1997) Cannibals With forks: The triple bottom line of 21st century business. Capstone Publishing Ltd, Oxford
Hofstede Insights (2020) Country comparisons. From https://www.hofstede-insights.com/country-comparison. Accessed 29 July 2023
Kerlinger F (1964) Foundations of behavioral research. Holt Rinehart & Winston, New York
Mack N, Woodsong S, MacQueen K, Guest G, Namey N (2005) Qualitative research methods overview: a data collector's field guide. Family Health International, Research Triangle Park
Miska C, Szocs I, Schiffinger M (2018) Culture's effects on corporate sustainability practices: a multi-domain and multi-level view. J World Bus 53:263–279
Riessman CK (2008) Narrative methods for the human sciences. Sage, London
Robinson OC (2014) Sampling in interview-based qualitative research: a theoretical and practical guide. Qual Res Psychol 11(1):25–41
Singh J (1984) Education and development: operative and espoused views of the school's role in Guyana. J Can Int Educ 13(2):9–21

2 Sustainability: Origin and Evolution

Abstract

This chapter begins with a restatement of the definition of sustainability first used in the Brundtland Report and then examines the three pillars of sustainability as proposed by John Elkington. These pillars, economic, environmental, and social are then examined in detail. In examining the economic pillar, the focus is on income inequality with data from various international organizations used to illustrate how income and wealth are distributed in various parts of the world. Analysis of the environmental pillar focuses on the degradation of various aspects of the environment with data from UNEP and other agencies used to illustrate the environmental crisis facing humanity. The discussion on the social pillar of sustainability focuses on the status of women globally.

As alluded to earlier, sustainability in its current usage originated in the 1987 report of the United Nations World Commission on Environment and Development (WCED). The simple definition suggested in the Brundtland Report as it is commonly known is that "sustainable development is the kind of development that meets the needs of the present without compromising the ability of future generations to meet their own needs." While the Brundtland definition focuses on the environmental aspects of sustainability the report tied the protection of the environment to social and economic justice. In the business literature sustainability is now widely seen to be achieved through Economic, Social and Governance (ESG) measures. This approach is similar to the triple bottom line that encompasses three pillars: *People, Planet, and Profit*. The term "the triple bottom line" was coined in 1994 by John Elkington.

> His (Elkington's) argument was that companies should be preparing three different (and quite separate) bottom lines. One is the traditional measure of corporate profit—the 'bottom line' of the profit and loss account. The second is the bottom line of a company's 'people account'—a measure in some shape or form of how socially responsible an organization has been throughout its operations. The third is the bottom line of the company's 'planet' account—a measure of how environmentally responsible it has been (Hindle 2009).

Elkington's unit of analysis was a company, but the three pillars approach is now used to assess sustainability in various other units, including countries. This process is fraught with ambiguities but is useful in determining progress that goes beyond simple GDP growth or per capita income. However, Elkington's framework is not universally accepted.

J. Singh, *Sustainability and the Sustainable Development Goals in Cultural Clusters*, Sustainable Development Goals Series, https://doi.org/10.1007/978-3-031-70618-9_2

2.1 Criticism of Elkington

Perhaps the most telling criticism of Elkington's "triple bottom line" is from the author himself. In a June 2018 article in the *Harvard Business Review*, he cites examples of product recalls to fix defects to argue for a voluntary recall of his TBL (triple bottom line) concept, presumably to address problems with it. "I am volunteering to carry out a management concept recall: With 2019 marking the 25th anniversary of the 'triple bottom line', a term I coined in 1994, I propose a strategic recall to do some fine tuning" (Elkington 2018).

The TBL concept has attracted much attention, so it is somewhat surprising that Elkington sees problems. However, it appears as if the main problem with the concept is that it is overambitious. Elkington's framework, which influenced initiatives like the Global Reporting Initiative and the Dow Jones Sustainability Indexes, was intended by him to "provoke deeper thinking about capitalism and its future, but many early adopters understood the concept as a balancing act, adopting a trade-off mentality" (Elkington 2018). Elkington is apparently disappointed that his concept has become an accounting system rather than an impetus to develop a new and sustainable capitalism.

Prior to Elkington's voluntary recall, his TBL concept—although receiving considerable attention and influencing a new approach to accountability in business—was criticized from different angles. Sridhar and Jones (2013) identify three such criticisms from the literature and seek to substantiate them by looking at 40 sustainability reports. The criticisms they identify are with measurement, aggregation, and TBL being a compliance mechanism. The first criticism on measurement is that it is relatively easy to measure financial performance, but more difficult to measure social and environmental performance. Moreover, as per the second criticism, there is no reliable method to aggregate measurements of the three pillars. This inability to aggregate results makes it difficult to compare performance to set targets or the performance of competitors. The third criticism suggests that corporations may aim for simply complying with indicators identified in sustainability indices, rather than doing as best as they could. In addition to these criticisms, it has been suggested that there are more than the three pillars covered in Elkington's concept. Burford et al. (2013) suggest that a fourth pillar could be "human values" and incorporate how cultural-aesthetic, political-institutional, and religious-spiritual considerations impact employees' personal and professional lives. A key consideration here may be whether the objective is to measure the impact of this dimension on organizations or vice-versa.

While Elkington has lamented the TBL becoming an accounting concept rather than a way to reform capitalism, the concept of sustainability has taken on new dimensions and is now seen as a way perhaps not to reform capitalism but to foster the development of a just world community. The United Nations Sustainability Development Goals (SDG) may be seen as a monumental effort in this direction. The 17 goals aim to achieve sustainability through economic, environmental, and social measures.

Economic Pillar of Sustainability The most significant barrier to economic sustainability is inequality within and among nations. Economic inequality has been engaging the attention of world leaders recently; not necessarily because they care deeply about the poor of the world but because vast disparities in income and wealth could threaten the stability of their countries and their positions within them. It is for this reason that business and political leaders gather annually for a World Economic Forum in Davos, Switzerland, to look at ways to mitigate the threat to their status inherent in an economically lopsided planet. Every year to mark the beginning of the Forum, Oxfam, the international charity group, issues a headline-grabbing statement illustrating global wealth inequality. On January 21, 2019, CNN reported on Oxfam's annual report issued to coincide with the beginning of the World Economic Forum. It highlighted a belief shared by many that there is no shortage of money in the world, but that it is not distributed fairly. According to Oxfam's 2019 report, 26 of the world's richest individuals have net assets equal to that of 3.8 billion of the world's poor, which is approximately half of the world's population (Luhby 2019). In its 2023 report, the organization shared that "the richest 1 percent grabbed nearly two-thirds of all new wealth worth $42 trillion created since 2020, almost twice as much money as the bottom 99 percent of the world's population. At the same time, at least 1.7 billion workers now live in countries where inflation is outpacing wages, and over 820 million people — roughly one in ten people on Earth—are going hungry" (Oxfam 2023). It is widely believed that this lopsided distribution of wealth is a systemic problem that is the result of deliberate policies promoted by some economists and politicians. Unchecked, many believe this will undoubtedly lead to revolution. Thus, there is an element of self-interest in the concern shown by the rich and powerful about the worrying issue of economic inequality. The global disparity in wealth as reported by Oxfam in 2019 and 2023 is graphically presented in Fig. 2.1. At the top of the pyramid, it can be seen that 1.2% of the world's population hold 47.8% of the world's wealth, while at the bottom, 53.2% percent hold just 1.1%.

Noted management scholar Peter Drucker is often credited with the famous business saying to the effect that "what can't be measured can't be managed." To properly study and perhaps manage income inequality, it is therefore important that valid and reliable indicators be employed. One of the most widely used indices of income inequality is the Gini coefficient. This coefficient estimates income distribution in countries according to how much it varies from a perfectly equal distribution. Therefore, in a hypothetically equal distribution, the Gini coefficient will be zero; while in a hypothetical case of maximum inequality; the coefficient will be 1 (a situation where one person gets all the income). The coefficient is named for its inventor, the late Corrado Gini, an Italian statistician. Gini built on the work of American economist Max Lorenz who, as shown in Fig. 2.2, illustrated income distribution by plotting cumulative income on the vertical axis against cumulative population on the horizontal. The resulting curve shows the deviation from perfect equality. The Gini coefficient is calculated by dividing the area A by the sum of areas A and B (A/A + B).

Comparing Gini coefficients by country and through time makes for interesting reading. As shown in Table 2.1, the Slovak Republic had the lowest Gini coefficient of 23.2, and the lowest 20 is rounded out by Poland with a coefficient of 28.8.

As shown in Table 2.2, South Africa had the highest reported Gini coefficient. It is interesting to note that all 20 countries with the highest Gini coefficients except St. Lucia, which is in the Caribbean, are either in Africa or Latin America. What might be the explanation for this?

In Table 2.3, the Gini coefficients for the five countries in this study are shown over time from 1979 to 2019. The lone number for Guyana indicates that in the late 1990s, it had the highest income inequality of the countries studied. Canada recorded the lowest level of income inequality overall with the coefficient remaining in the 33–34 range from 1997 to 2019. The reported readings for India in 2009 and 2019 were close to 35. Sri Lanka's Gini coefficient for those 2 years was 36.1 and 37.7 respectively. In the period from 1979 to 2019, the USA moved from 35 to 41. Thus, as shown in Table 2.3, Canada showed the lowest income inequality, and the USA the highest of the countries for which data is available over time.

Frederick Solt (2019), acknowledging that a problem could arise in comparing Gini data from multiple and diverse sources, standardized the coefficients to make cross-national comparisons valid. Solt's Standardized World Income Inequality Database (SWIID) is aimed at facilitating the comparison of income inequality data by country and over time. Figure 2.3 shows how standardized income inequality, as measured by disposable income in Solt's database, has unfolded in Canada, the USA, India, and Sri Lanka. The data in this graph are more complete than what is shown in Table 2.3 and indicate that

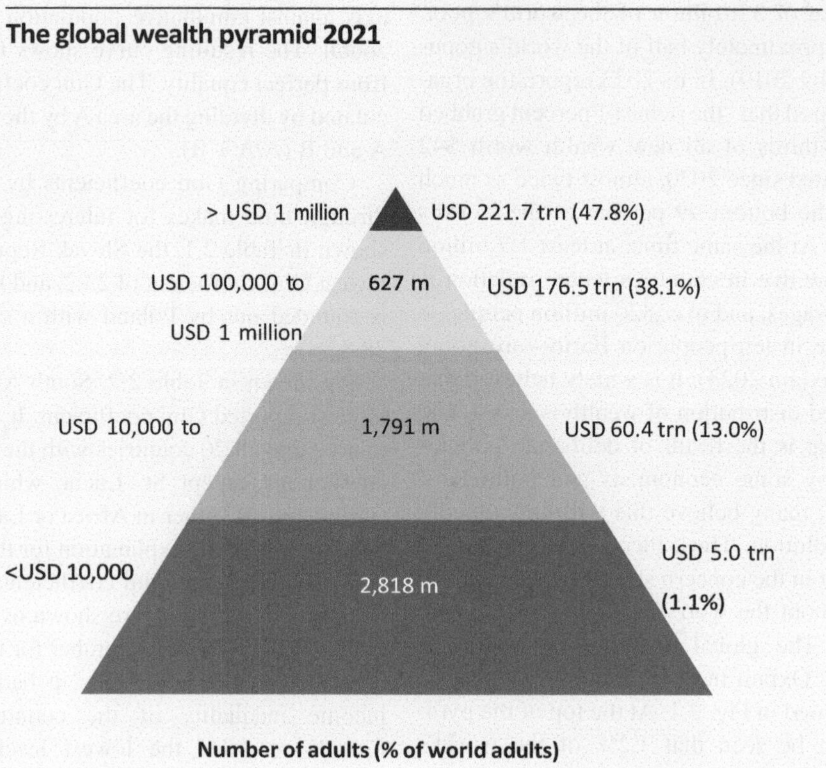

Fig. 2.1 The global wealth pyramid. (Davies et al. 2022)

Fig. 2.2 The Gini coefficient. (Haughton and Khandker 2009)

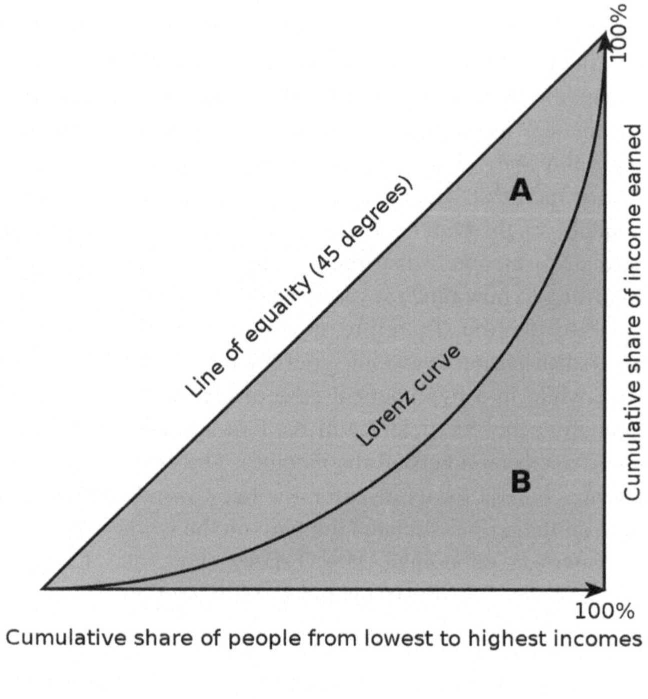

Table 2.1 Countries with the 20 lowest Gini coefficients

Country	Most recent year	Most recent value
Slovak Republic	2019	23.2
Slovenia	2020	24.0
Belarus	2020	24.4
Ukraine	2020	25.6
Moldova	2021	25.7
Belgium	2020	26.0
Netherlands	2020	26.0
United Arab Emirates	2018	26.0
Iceland	2017	26.1
Czechia	2020	26.2
Azerbaijan	2005	26.6
Finland	2020	27.1
Denmark	2020	27.5
Algeria	2011	27.6
Norway	2019	27.7
Kiribati	2019	27.8
Kazakhstan	2018	27.8
Armenia	2021	27.9
Timor-Leste	2014	28.7
Poland	2019	28.8

World Bank (2022)

Table 2.2 Countries with the 20 highest Gini coefficients

Country	Most recent year	Most recent value
South Africa	2014	63.0
Namibia	2015	59.1
Zambia	2015	57.1
Central African Republic	2008	56.2
Eswatini	2016	54.6
Mozambique	2014	54.0
Botswana	2015	53.3
Brazil	2021	52.9
Colombia	2021	51.5
Angola	2018	51.3
St. Lucia	2016	51.2
Panama	2021	50.9
Zimbabwe	2019	50.3
Congo, Rep.	2011	48.9
Costa Rica	2021	48.7
Guatemala	2014	48.3
Honduras	2019	48.2
Burkina Faso	2018	47.3
Cameroon	2014	46.6
Nicaragua	2014	46.2

World Bank (2022)

India has the most pronounced income inequality of the four countries in the study covered in SWIID.

The Gini coefficient as a measure of income inequality has its limitations. The first has to do with how income is measured: if income is measured at the individual level, the coefficient will be different than if it is measured at the household level. The Gini index also does not account for state programs aimed at alleviating poverty and thus mitigating some of the effects of income disparity. Furthermore, it does not account for the informal sector, which is a major part of the economies of developing countries. As noted by researchers in 2014, another problem with the index exists.

> Moreover, two countries could have different income distributions but the same Gini index. For example, in a country where 50% of the people have no income and the other 50% of the people have equal income, the Gini index is 0.5. In another scenario, where 75% of people with no income account for 25% of a country's total income, and the top 25% of people with an income account for 75% of the country's total income, the Gini index will also be 0.5. Consequently, as a basis for ranking the differences in income inequality between countries, the Gini index could be misleading (Chitiga et al. 2014).

As a result of the many limitations of using the Gini coefficient as an indicator of income inequality, other measures have been developed to analyze this important aspect of sustainability. One such approach is to track the percentage of national income flowing to the top 10% of the population. As shown in Fig. 2.4, this percentage illustrating the lopsided distribution of income globally rose to almost 58% in 2000 and then fell back to just over 52% in 2020. Is such disparity sustainable?

In Fig. 2.4, the percentages of national income flowing to the top 10% in the countries covered by this study are shown. The highest amount of national income flowing to the top 10% is in India. This figure, which was just over 30% in 1980, was on an upward trend until flattening out at about 56%. Impressive economic growth in India has therefore been accompanied by worsening income inequality. The estimate for Guyana

Table 2.3 Gini coefficients over time

Gini index by country and year							
	1979	1989	1997	1998	1999	2009	2019
Canada	–	–	33.1	33.2	33.1	34.0	–
Guyana	–	–	–	45.1	–	–	–
India	–	–	–	–	–	35.4	35.7
Sri Lanka	–	–	–	–	–	36.1	37.7
United States	35.3	38.2	40.5	40.0	40.0	40.6	41.5

World Bank, World Development Indicators
Created from: World Development Indicators Series: Gini index

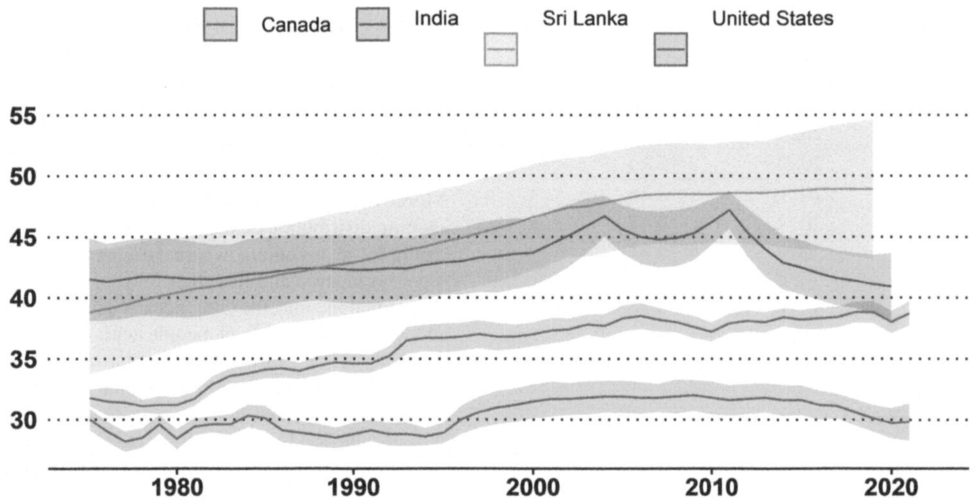

Fig. 2.3 SWIID Gini index over time

is also above 50%. The trend in all five of the countries shown in Fig. 2.5 is upward signaling worsening income inequality. Again, the question that arises is whether such income disparity is sustainable.

Figure 2.6 shows what percentage of national income is received by the top 10% of the population in a sample of regions. In 2023, the top 10% in Europe received approximately 37% of national income, while in the other regions shown the share of income flowing to the top earners has been moderating from well above 50%. The trend in North America is on the rise towards the 50% mark.

Figure 2.7 shows the share of national income received by the richest 1% in the said sample of regions. As shown in the figure, the top 1% globally received just under 20% of total income. The pattern of income distribution shown for the top 1% is similar to that for the top 10%. In both cases, the inference is one of severe income inequality. As will be discussed later, some Indian interviewees in this study rationalized this disparity by suggesting that those at the bottom were also doing better than before and that they were now able to acquire goods that were beyond their reach before economic liberalization. It should be noted that in the US-Canada region, the disproportionate amount of income flowing to those in the top 1% rose from just over 10% to 20% in the more than 40 years of data charted in Fig. 2.6. As will be seen in the analysis of the interviews with political and business leaders in Canada and the USA, this may lead to social challenges.

Obviously, if disproportionate income is flowing to those at the top, there will be less flowing to those at the bottom. This is manifested in the data presented in Fig. 2.8 showing the share of

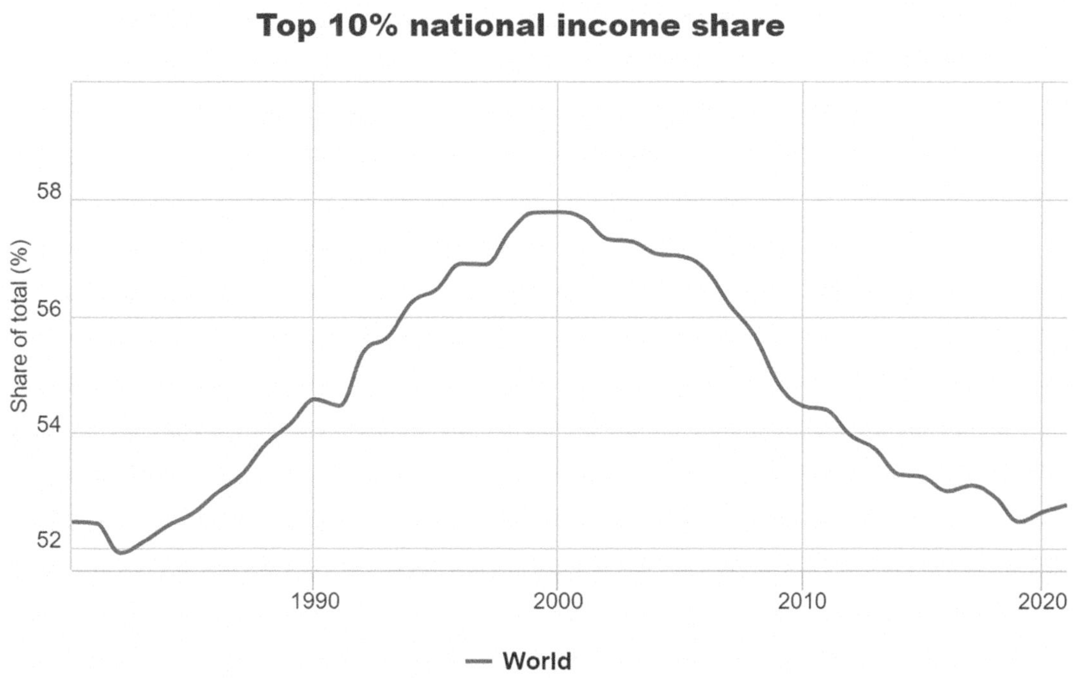

Fig. 2.4 Share of income received by top 10% globally. (World Inequality Database 2023)

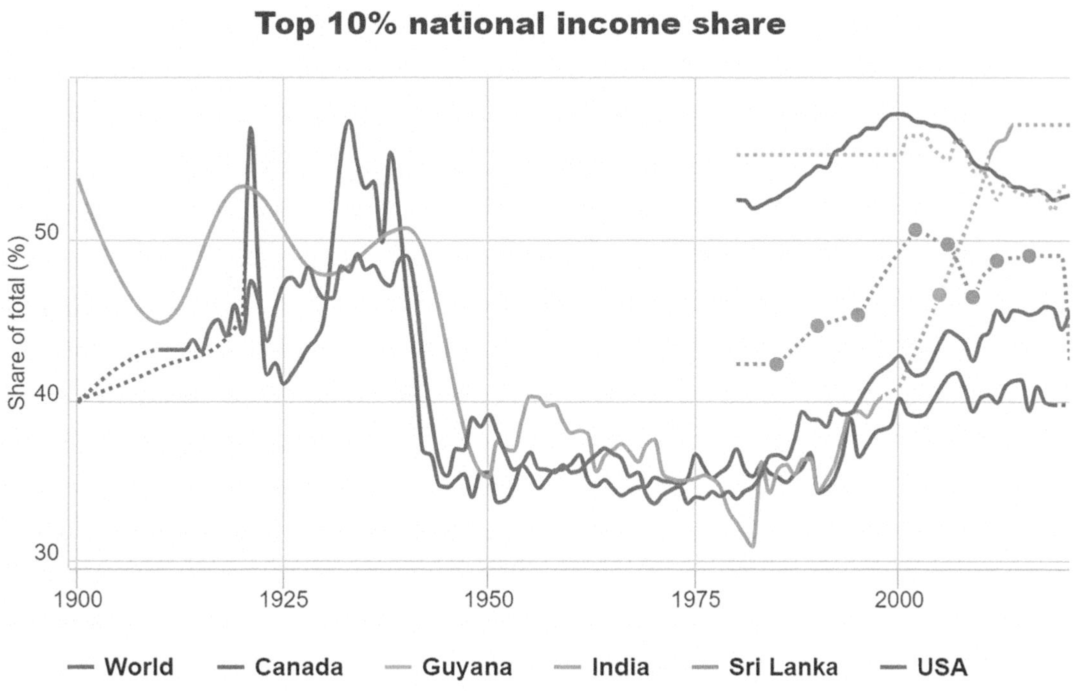

Fig. 2.5 Share of national income received by top 10% (I)

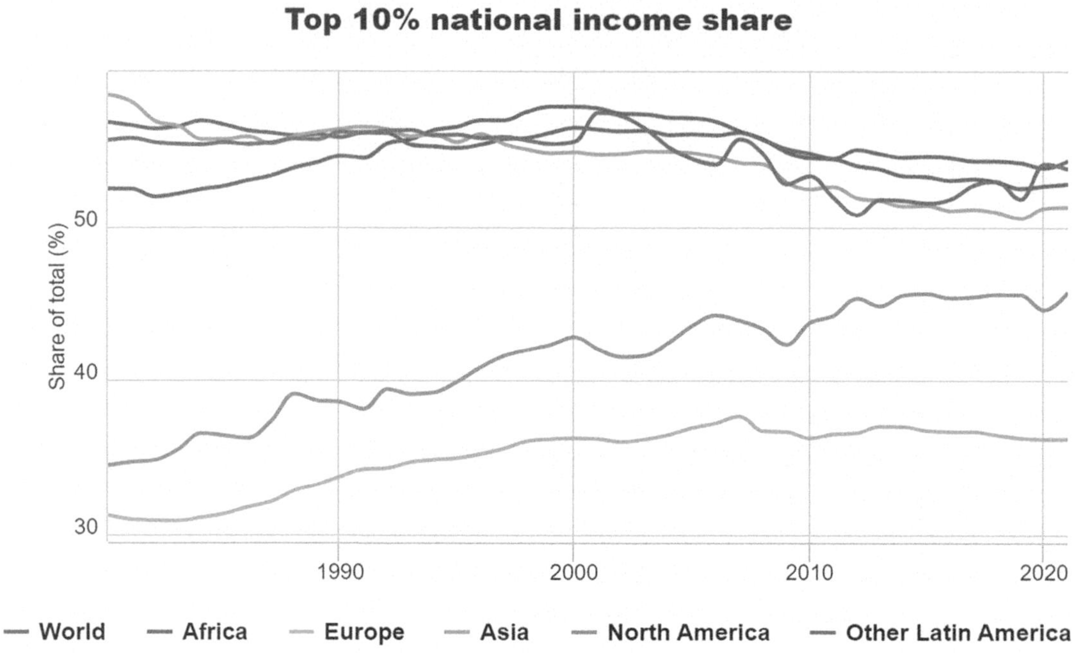

Fig. 2.6 Share of national income received by top 10% (II)

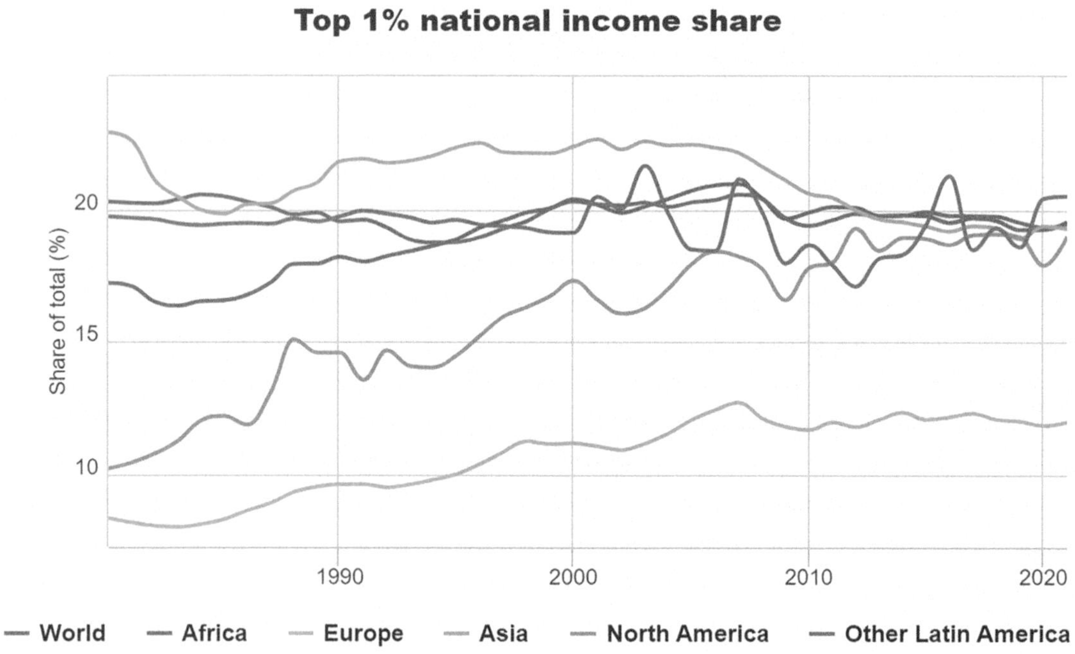

Fig. 2.7 Share of national income received by top 1%

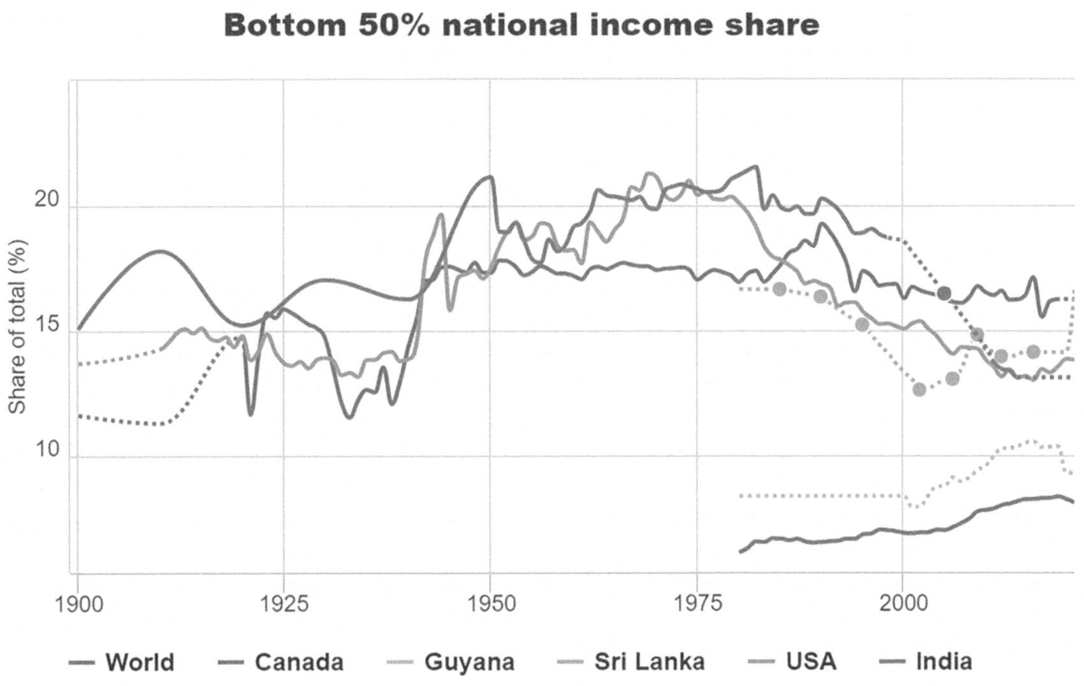

Fig. 2.8 Share of national income received by bottom 50%

income received by the bottom 50% in the countries where interviews were conducted for this study. This number was lowest in Guyana and trending lower in the other countries. As the income and wealth disparity grows, political leaders will face a massive problem in containing rising discontent and unrest.

However, one factor that may be stemming the tide towards such discontent is an overall rise in income and wealth around the world. As it relates to income, in the period from 1980 to 2016, the average income for the world rose 60% as measured in Purchasing Power Parity (PPP Euros) net of inflation. However, the increase was uneven among countries, with China seeing an increase of 831% and Russia just 34%. Moreover, the general trend has been for the increases at the top to be substantially higher than those at the bottom (WID.world, 2017). So, the overall trend in income growth is that there has been an increase in the size of the pie—but those at the top received a disproportionate share of it. This is in sync with trickle-down economic policies aimed at reducing taxes for higher brackets and businesses, together with fewer regulations. The expectation is that such policies will lead to higher incomes at the top and also at the bottom. However, as the figures indicate, the rate of increase at the top far outstrips those at the bottom, leading to an exacerbation of economic disparity.

The extent of this growing income disparity worldwide is captured in Fig. 2.9 where the ratio of income of the top 10 to bottom 50 is shown over time. This ratio has fluctuated over time, falling from a high of more than 40 in 1980 to around 22 in 2020. This means that throughout the period for which data is reported in the World Inequality Database, the income of just the top 10% was consistently significantly higher than that of the bottom 50%.

2.2 Environmental Sustainability

The Summer of 2023 will be remembered for massive forest fires and severe storms that had global reach. The smoke from Canada's forest

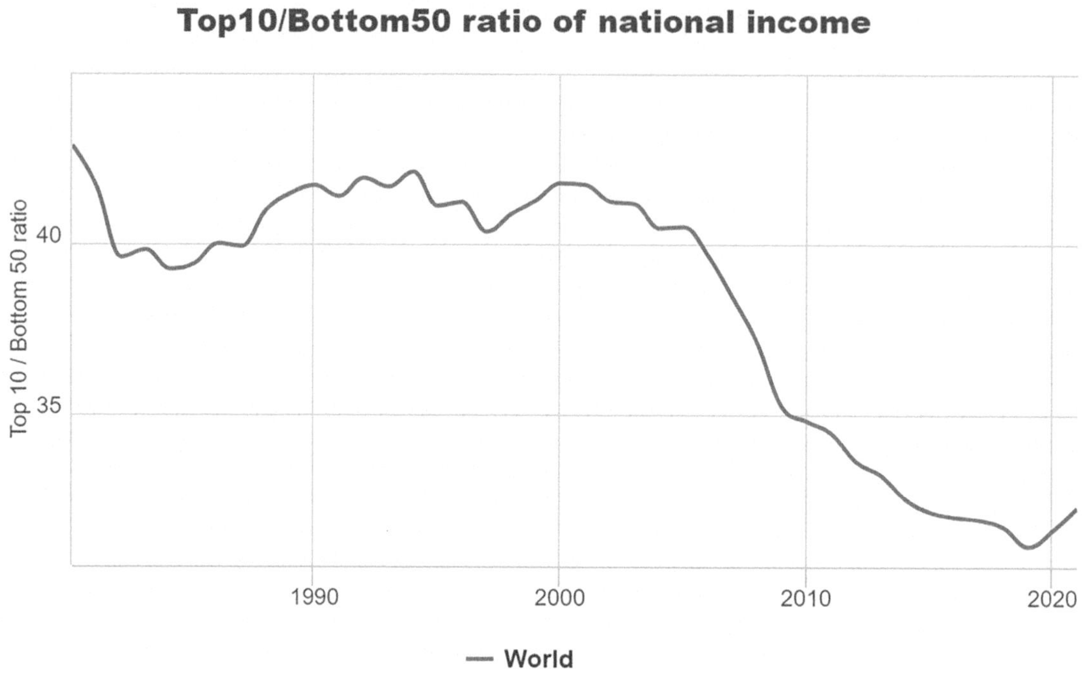

Fig. 2.9 Top 10% compared to bottom 50% globally

fires reached as far as Europe and was a reminder of the interconnectedness of the planet's environment. The severity of the fires and storms is attributable to global warming. Trickle-down economic policies, accompanied by the rollback of environmental regulations, have created a situation that some see as an existential threat to humanity. This is addressed in the environmental pillar of sustainability. Can we continue current production modes and consumption levels if the planet is to survive? Many believe that urgent action is needed to stem the tide of environmental degradation. However, there are powerful political actors around the globe who still believe that the environmental threat to humanity is a hoax. Former US president Donald Trump is foremost among these politicians, but he is not alone. Former Brazilian President Jair Bolsonaro is another world leader who has questioned the need for any action on the environment and rolled back environmental regulations in his country. His actions on environmental issues have been so callous that he has been labelled as "arguably the most environmentally dangerous head of state in the world" (*The Amazon is Approaching*, August 3, 2019, p. 14). The Amazon basin contains 40% of the globe's tropical forests and is extremely important because of its biodiversity and the role it plays as a sink for carbon dioxide. Just over 60% of the Amazon rainforest lies within Brazilian territory, so the government of Brazil has a crucial role to play in preserving an area that has been described as the lungs of the earth. However, President Bolsonaro did not seem to appreciate the significance of the Amazon rainforest to the world's ecology. His successor, President Lula da Silva has taken measures to stop the destruction of this vitally important ecological area, but the damage already done will be difficult to repair.

Neighboring Guyana, where some of the interviews for this study were conducted, controls just 3% of the Amazon rainforest but has been a model world citizen in preserving its part of the "lungs of the earth." Growing out an offer made by Guyana to the Commonwealth in 1989, the Iwokrama Rainforest Programme is aimed at, among other things, preserving the biodiversity of the rainforest. The Government of Guyana has set aside 371,000 ha (3710 km^2) of forest—about 1.6% of Guyana's landmass—for this endeavor. Bolsonaro, on the other hand, since his election in 2018, had embarked on a process to dismantle or frustrate the enforcement of environmental protection legislation. His environment minister, Ricardo Salles, was quoted as saying, "The role of the state is to protect landowners' property rights," and wanted to use donated funds from Norway and Germany to compensate landowners for conservation projects carved out of areas largely illegally occupied (*The Amazon is Approaching*, August 3, 2019). *The Economist* estimates that during Bolsonaro's tenure, at least roughly 4300 km^2 of Brazil's rainforest had been lost. Brazil is not the only country where the degradation of the environment is of grave concern. It is a global issue that not only concerns the forests but every aspect of the global ecosystem.

What is the state of the global environment? The United Nations Environment Programme provides a periodic assessment of the planet's environment, which evaluates the state of five environmental elements: air, biodiversity, oceans and coasts, land and soil, and freshwater. GEO-6, as the most recent report is known, claims that "a healthy environment is the best foundation for economic prosperity, human health, and well-being" (UNEP 2019, p. 4). However, the report found that human behavior has negatively impacted the five elements and the resulting environmental degradation, ranging from serious to irreversible, has harmed human health. The linkages between the drivers and responses of atmospheric change are shown in Fig. 2.10.

Geo-6 states that emissions caused by human activity have caused changes in the atmosphere in the areas of air pollution, climate change, stratospheric ozone depletion, and bioaccumulative toxic substances (UNEP 2019). These, in turn, have an impact on the health of the planet and its inhabitants. The linkages between human activity and their impacts are shown in Fig. 2.8. In its 2019 report, UNEP found that concentrations of carbon dioxide and other greenhouse gases continue to increase and attributes this mainly to the use of fossil fuels in energy generation. This leads to global warming, which in turn leads to, among other things, rising sea levels. The report further declares that: "There is robust evidence that climate change and increased climate variability worsen existing poverty, exacerbate inequalities, and trigger new vulnerabilities" (UNEP 2019, p. 108), and warns things will only get worse if measures are not taken to stop greenhouse gas emissions. However, the report found a bright spot in the reduced emission of ozone-depleting chemicals and attributes that to the Montreal Protocol, hence suggesting that international agreements can be effective in curbing the emission of particular harmful chemicals.

The increased emission of greenhouse gases is attributed mainly to the burning of fossil fuels in the generation of electricity. As shown in Fig. 2.11, the bulk of fuel used in electricity generation around the world is fossil fuels, with coal, oil, and natural gas making up approximately 61% of total consumption. As mentioned earlier, emissions harm human health in addition to threatening food security and ecosystems. UNEP (2019) suggests that air pollution was responsible for 6–7 million premature deaths worldwide in 2016. Additionally, the report cited a WHO study that found long-term exposure to particulate matter, a category of emissions, was responsible for between 3.6 and 4.6 million of those deaths, and millions of otherwise healthy lives were lost to diseases such as lung cancer and heart disease. Overall, therefore, the state of the planet's air is dismal and will get worse unless drastic measures are taken to curb greenhouse gases.

The increased emission of greenhouse gases has led to global warming that is causing climate change. This in turn is affecting biodiversity, which is one of the five elements of the environment covered by the UNEP 2019 report. Climate change has caused significant changes in biodiversity as vividly manifested in recent Australian forest fires. These massive fires have become frequent occurrences in Australia, with the 2020 fire affecting every state in the country and killing 28 people. According to reports, as much as 7.3 million hectares burned, and as many as 1 billion animals were affected. Remarkably, it is estimated that almost a third of the koalas in New

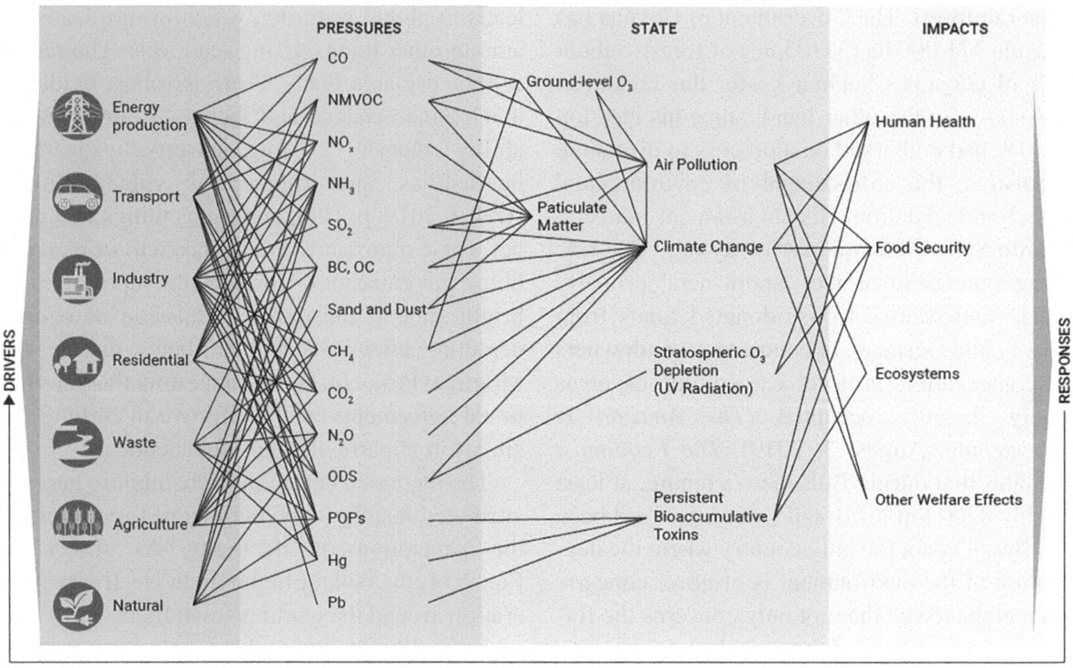

Fig. 2.10 Primary linkages among pressures, state, and impacts of atmospheric change. (UNEP 2019, p. 109)

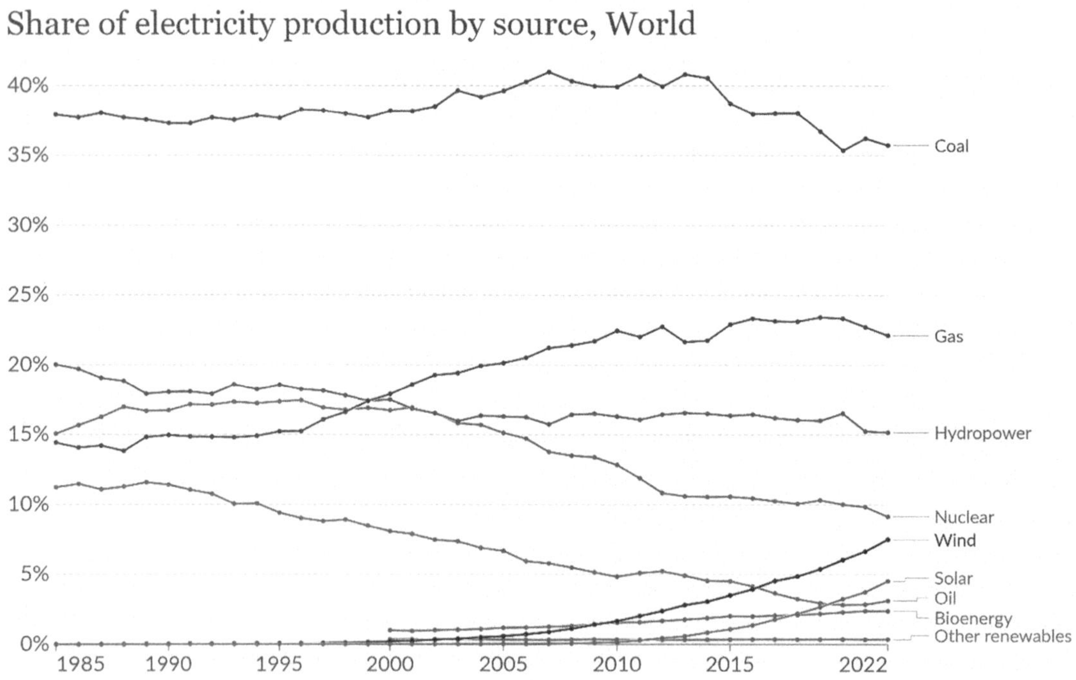

Fig. 2.11 World electricity generation by source. (Our World in Data 2023)

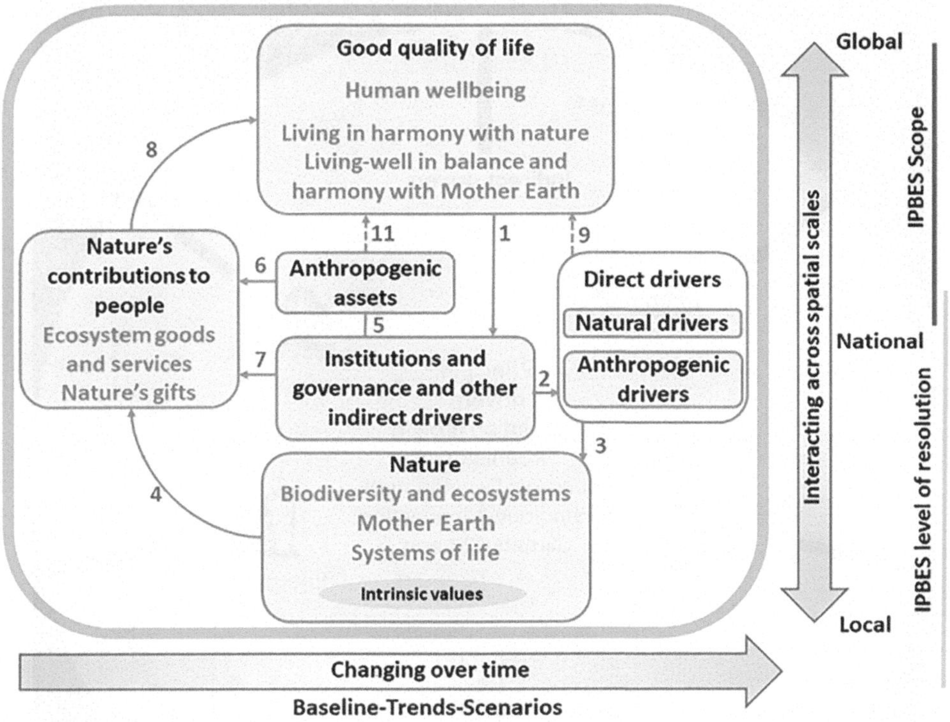

Fig. 2.12 The relationship between biodiversity and sustainable development. (Diaz et al., 2015, p.5)

South Wales were killed by the 2019–2020 fires (Jessie Yeung, CNN, January 13, 2020). The damage to biodiversity caused by forest fires is only part of the destruction of plants and animals worldwide.

According to UNEP's 2019 report, biodiversity is in serious trouble with there being strong evidence of continuing reduction in species diversity. The report suggests scientists are worried that if current anthropogenic pressures on biodiversity continue "we risk precipitating a sixth mass extinction event in Earth history" (UNEP 2019, p. 142). As shown in Fig. 2.12, the decline in species diversity could have an impact on the quality of life of people worldwide.

Similarly, the health of the planet's oceans has an impact on people's lives. UNEP found that all is not well with oceans and coasts, as a growing population puts pressure on the health of this important sector of the global environment. Overall, ocean levels are rising because of global warming, and there are other areas of concern—such as the deterioration of coral reefs, which are a critical part of the ocean ecosystem. As widely reported in late March 2020, Australia's Great Barrier Reef was experiencing its most severe bleaching ever. This process, which involves the expulsion of algae living in the corals, leads to the corals becoming white. Although the corals don't automatically die because of this, it places severe stress on the coral reefs. This bleaching is a direct result of rising ocean temperatures caused by global warming.

Another factor putting pressure on the health of oceans is the enormous amount of litter that has accumulated in them. According to the UNEP 2019 report, an estimated 8 million tons of plastics find their way into oceans annually, and if steps are not taken to stanch the flow, the amount of plastics in the ocean could rise to between 100 and 250 million tons by 2025. The main cause of the alarming rise of plastics in the ocean is the mismanagement of waste in coastal areas. As is well known, some of these plastics find their way

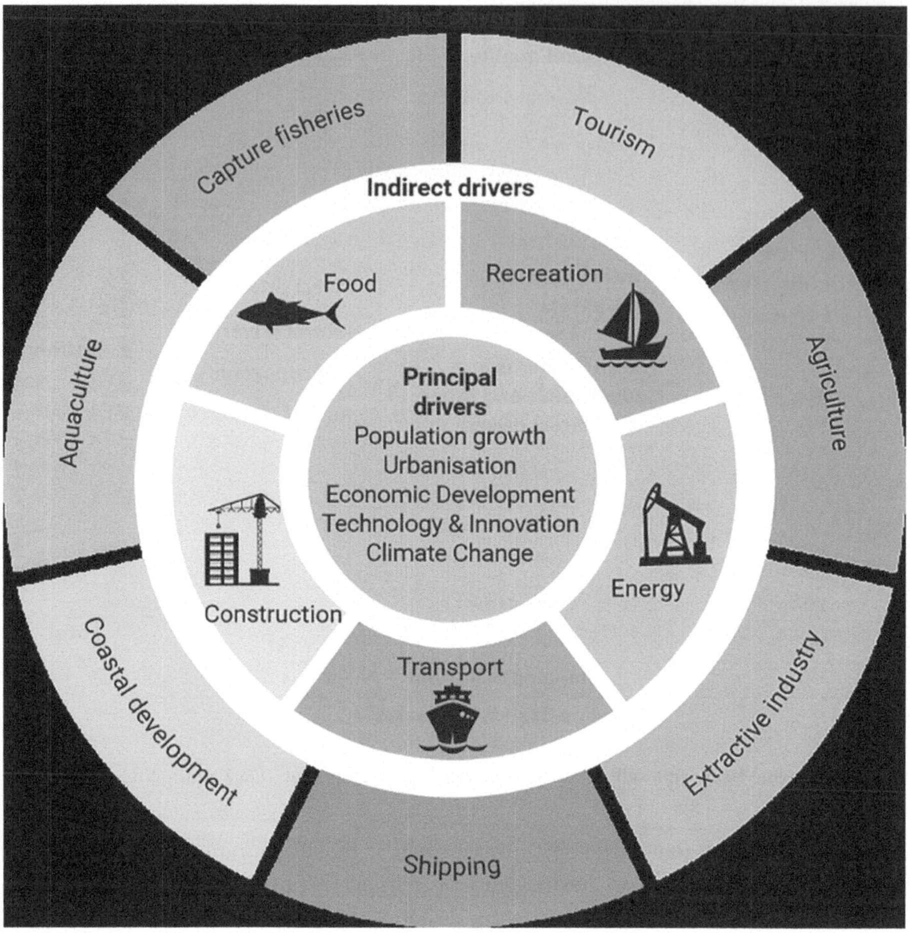

Fig. 2.13 Oceans, coasts, and socioeconomic drivers

into our food supply. The overall link between oceans and coasts and socioeconomic drivers is shown in Fig. 2.13 (UNEP 2019, p. 179).

As with oceans and coasts, UNEP (2019) found the land and soil segment of the environment to be in a poor state and under increasing pressure. The report found that food production to support a growing world population accounted for 50% of habitable land in the world. A notable finding of this report is that while this was the foremost anthropogenic use of land, roughly one-third of global food production is lost or wasted with slightly more than half of this waste occurring in industrialized countries. If this loss of wasted food could be eliminated, there would be less need for agricultural land. Nevertheless, it is projected that even with higher efficiency and better agricultural practices, it will be difficult to produce enough food for the expected world population of 10 billion in 2050. The challenge of producing enough food for a growing world population will undoubtedly lead to conflict. A foretaste of this is the Syrian conflict which many have blamed on environmental factors. Syria today is hardly a functioning country, and large parts of it have been destroyed by a war that has its roots in a severe drought between 2007 and 2010 (Kelley et al. 2015). That drought caused extensive crop failure that led to some 1.5 million people moving closer to urban areas. This migration placed severe pressure on government services, which culminated in an upheaval that eventually saw two-thirds of the country's 22 million people displaced. Figure 2.14 shows the

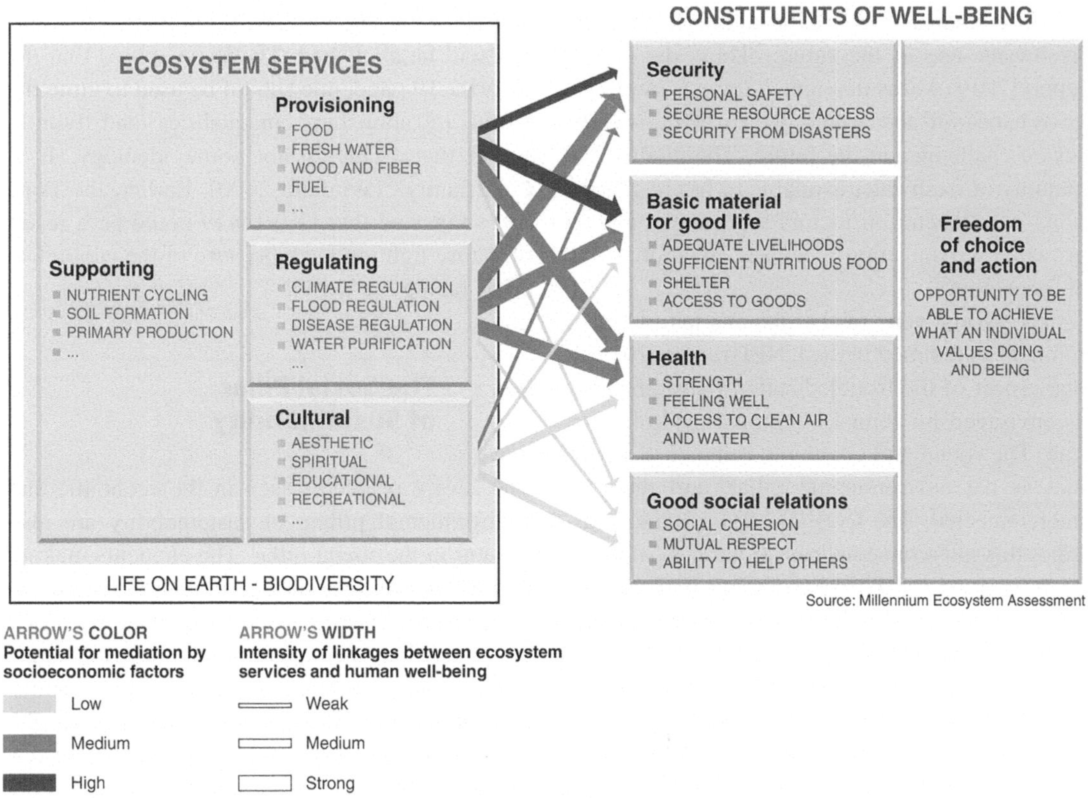

Fig. 2.14 The ecosystem and human well-being. (Millennium Ecosystem Assessment 2005, p. 227)

linkages that, when broken by land and soil degradation, can lead to upheavals such as the one in Syria.

As with forest fires and intense weather events, freshwater shortage has become a common occurrence around the world. Although 70% of planet Earth is covered with water, there is a growing shortage of fresh water that people can use in their daily lives to drink, clean themselves, and/or irrigate their fields. Freshwater accounts for only 3% of the world's water, and two-thirds of that is unavailable for use by humans. This, coupled with a growing world population, has led to billions of people not having adequate access to water (WWF 2020).

Countries such as India, Iran, and Botswana are among 17 countries that are regarded as being under high water stress. According to the World Resources Institute, this means these countries have no excess water. Some of these countries are dependent on groundwater, which experts suggest only be used during droughts. Moreover, the use of groundwater literally leads to the sinking of locales. Within those countries are several big, thirsty cities that have faced acute shortages recently, including São Paulo, Brazil; Chennai, India; and Cape Town, South Africa which in 2018 narrowly beat what it called Day Zero—the day when all its dams would be dry (New York Times, August 6, 2019).

Cape Town was expected to run out of water on July 9, 2018, but Day Zero was averted by adequate rainfall that arrived earlier than expected. If Day Zero had not been averted, Cape Town would have become the first major city in the world to run out of water. The water shortage in Cape Town was caused by a prolonged drought and averted by strict conservation and timely rainfall.

Unless something changes, many other cities could face a similar situation in the future. This will add to several other expected problems with freshwater use in the future. Thus, the UNEP report (2019) warns that unless there is improved governance of the water cycle, there will be severe challenges in the future. The quality and quantity of freshwater available to future generations will depend on factors such as population growth, urbanization, industrialization, and deforestation.

The dismal state of the planet's natural environment as outlined in the UNEP report (2019) is one aspect of the troubled state of sustainability as envisaged by John Elkington's triple bottom line. The social and economic pillars are as dismal as the environmental pillar, and they are interconnected. The COVID-19 pandemic illustrates this interconnectedness (Quinney 2020).

Writing on the World Economic Forum platform at a time when almost three million people worldwide were infected with the COVID-19 virus (of which almost 250,000 had died), Quinney (2020) suggested that while investigators looked for the origin of the virus that the connection to the destruction of the natural habitat of wildlife cannot be discounted. She argued that "60% of infectious diseases originate from animals, and 70% of emerging infectious diseases originate from wildlife. AIDS, for example, came from chimpanzees, and SARS is thought to have been transmitted from an animal still unknown to this day. We have lost 60% of all wildlife in the last 50 years, while the number of new infectious diseases has quadrupled in the last 60 years. It is no coincidence that the destruction of ecosystems has coincided with a sharp increase in such diseases."

In turn, COVID-19 caused unprecedented disruption of social and economic activities around the world. Moreover, it has brought social and economic inequities to the fore. For example, in the United States, it was found that a disproportionate number of those dying from the infection were Black and poor (Kendi 2020). The interconnectedness of the pillars of sustainability illustrated by the COVID-19 tragedy has attracted strong comments from many world leaders and significantly from His Holiness Pope Francis, who in looking for a silver lining to the pandemic anticipated a post-virus world in which inequalities will be abolished. The Pope argued that the COVID-19 pandemic should be used to unite the world in abolishing inequalities and bluntly stated that "This is not some ideology. It is Christianity" (Winfield 2020). Earlier, the Pope had suggested that COVID-19 could be a result of nature fighting back because of the ecological crisis facing the world.

2.3 The Social Pillar of Sustainability

The severe problems seen in the economic and environmental pillars of sustainability are also present in the social pillar. The elements making up the social pillar, such as racial inequality and caste discrimination, vary more widely around the globe, but one constant that could provide a guide to the state of the social pillar is the status of women worldwide. Figure 2.15 illustrates the state of gender equality over time in various regions of the world. The trend is towards more equality with the notable exception of Eastern Europe which experienced a decline in the 1980s.

Klasen (2020) assessed this indicator of social equality and found a mixed bag suggesting that in Europe and North America, women's rights improved considerably in the twentieth century and many remaining areas of gender discrimination were erased after the Second World War. However, in the developing world, such progress was not realized until the attainment of political independence in the1940s to the 1960s. He thus concluded that: "When surveying the world from 1960 to the early years of the new millennium, progress on reducing gender inequality was widespread, fast, steady, and seemingly irreversible. This was particularly the case in developed countries, but there were indications that it also applied to most developing countries" (Klasen 2020, p. 2).

As shown in Fig. 2.16, the progress in reducing gender inequality is reflected in a worldwide reduction in the percentage of potential constraints placed on women's rights from 1960 to 2010. For example, in Sub-Saharan Africa (SSA),

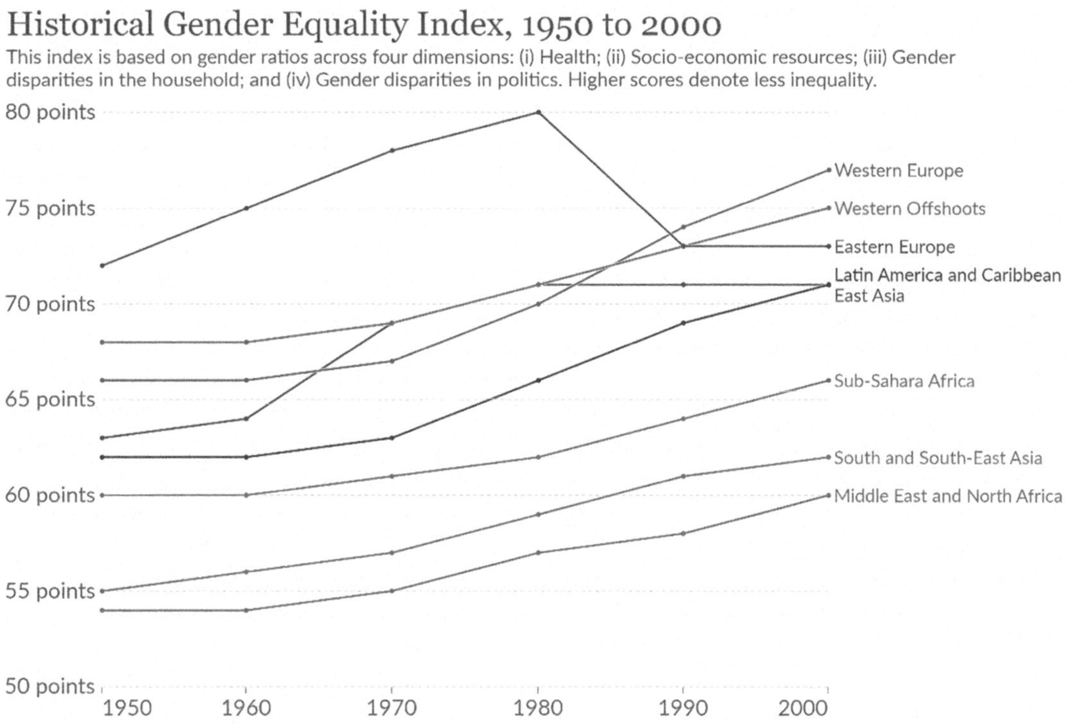

Fig. 2.15 Gender equality over time

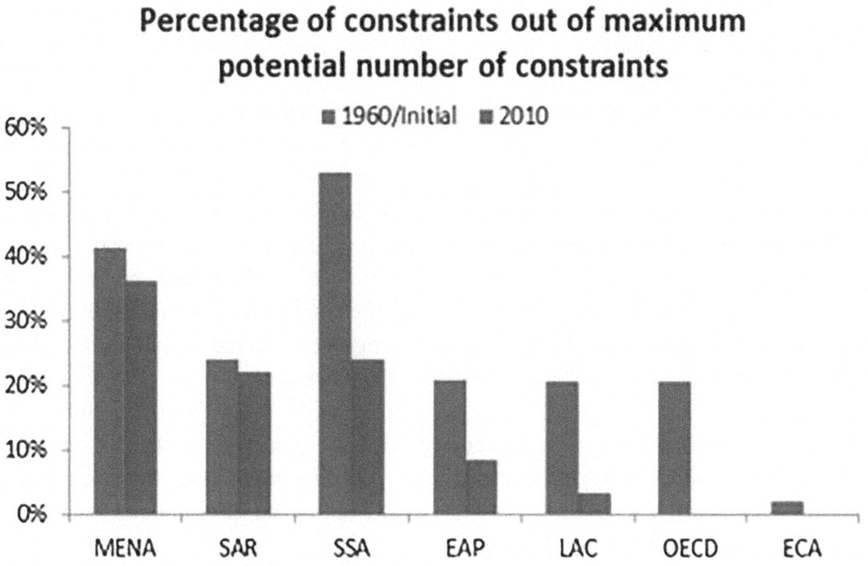

Fig. 2.16 Constraints on women's rights. *EAP* East Asia and Pacific, *ECA* Europe and Central Asia, *LAC* Latin America and Caribbean, *MENA* Middle East and North Africa, *OECD* Organization for Economic Cooperation and Development, *SAR* South Asia, *SSA* Sub-Saharan Africa. (Hallward-Driemer et al. 2013, as cited by Klasen 2020, p. 2)

the percentage of potential constraints fell from 50% to 20%. This does not mean that there are no constraints on women's rights in SSA, but there has been a marked reduction in such constraints. However, not much progress was made in the Middle East and North Africa (MENA) and Special Administrative Regions (SAR) of China in the same period. In the Organization for Economic Co-operation and Development (OECD) countries and Europe and Central Asia (ECA) constraints on women's rights have virtually been eliminated.

Another area in which considerable progress towards gender equality has been made is in education. As shown in Fig. 2.17, in OECD countries, women have achieved parity with men in secondary education since 1980. Moreover, from 1990, they outnumbered men in tertiary institutions. A similar trend is seen for the World Bank-defined Europe and Central Asia region that includes countries like the Russian Federation and Ukraine. However, such progress has not been realized in all parts of the world. For example, in Sub-Saharan Africa, fewer women than men are enrolled in both secondary and tertiary educational institutions.

Klasen (2020) also identified decreased gender inequality in mortality, noting that it had disappeared in developed countries by the early twentieth century and had been considerably reduced in developing countries. He concluded that: "The share of 'missing females' who had died as a result of gender inequality in mortality, fell from 7.9% around 1990 to 6.8% around 2000 in the regions most affected by it, with a decline recorded in all regions and countries except China" (Klasen 2020, p. 3). Many analysts have noted that the female population of the world is considerably lower than predicted by statistical models and attribute this to a preference for male children in some parts of the world and inequality in health care.

Accompanying improvements in rights, education, and health, were higher labor force participation rates for women. This signals a greater role for women in economic development.

The ratio of women to men's labor force participation rate over time is shown in Fig. 2.18.

The ratio worldwide has hovered around 70% from the 1990s to the present. This means that women still are not participating in the workforce at the same rate as men. Moreover, the rates at which they are participating represent a mixed bag, with initial improvements being followed by declines in South Asia, and stagnation in participation rates in many other regions. Klasen (2020) sees similar signs of stagnation, regress, and backlash in recent developments on gender equality and offers several reasons for their emergence. First, in many developed countries, female participation rates in the labor market are still lower than the male rate, and a large proportion of women are working in female-dominated occupations where wages are lower than in male-dominated occupations. Moreover, women are underrepresented and underpaid in high-level jobs, pointing to a glass ceiling effect.

This is reflected in female labor income share as shown in Fig. 2.19. While this proportion has been rising steadily since the 1990s it is still only about 35%.

Second, Klasen (2020) suggests that women in the workforce are often still the primary providers of care for their children and parents, shouldering the bulk of the responsibilities of the sandwich generation. The clear implication here is that in addition to the physical and emotional strain caused by such heavy responsibilities, they impede career progress.

Third, he suggests that, as illustrated by the recent prosecution of powerful men—such as Hollywood producer Harvey Weinstein—and the MeToo movement, some men in positions of power engage in the sexual harassment of women in the workplace.

Fourth, Klasen (2020) posits that globalization has engendered a shift from manufacturing to services, and this has benefitted women. However, at the same time, this has caused a loss of employment opportunities for skilled men. The modest gains made by women have therefore led to resentment and backlash from men disadvantaged by the effects of globalization and other socioeconomic factors. In the developing world, the substantial improvements in equality

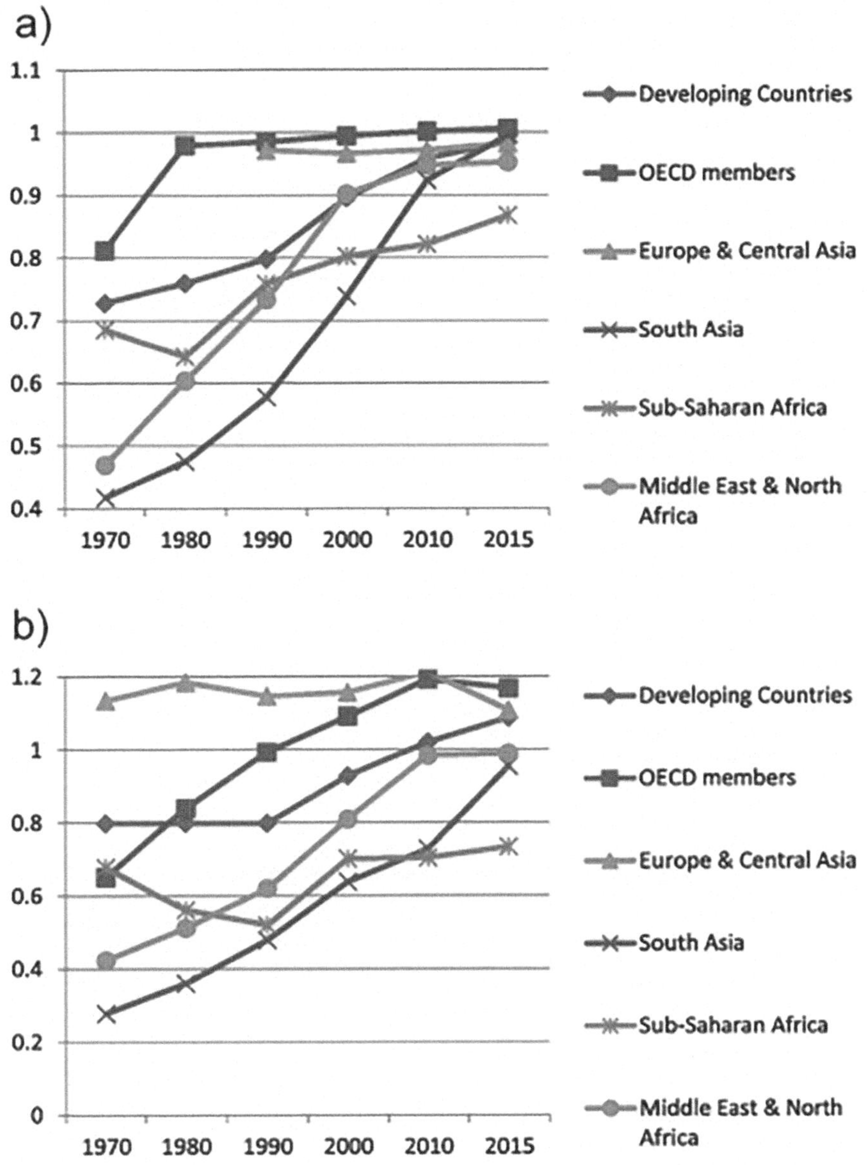

Fig. 2.17 Educational enrolment by gender. (**a**) Female-male ratio of secondary enrolment. (**b**) Female-male ratio of tertiary enrolment. Note: Developing countries are defined as being members of the World Bank's IBRD or IDA funding arms which includes nearly all non-OECD countries except some oil-rich Arab States (and Cuba and North Korea). In Europe and Central Asia and the Middle East and North Africa, only developing countries as defined above are included. OECD countries include Mexico and Turkey, which contributes to the low female-male ratios in 1970 and 1980. Source: World Development Indicators. (Klasen 2020, p. 2)

rights and education have not led to a concomitant reduction in female inequality in labor markets. Female labor force participation rates in developing countries generally continue to lag that of men and so have their wages for comparable work.

As shown immediately above, while the status of women globally has improved, there remain

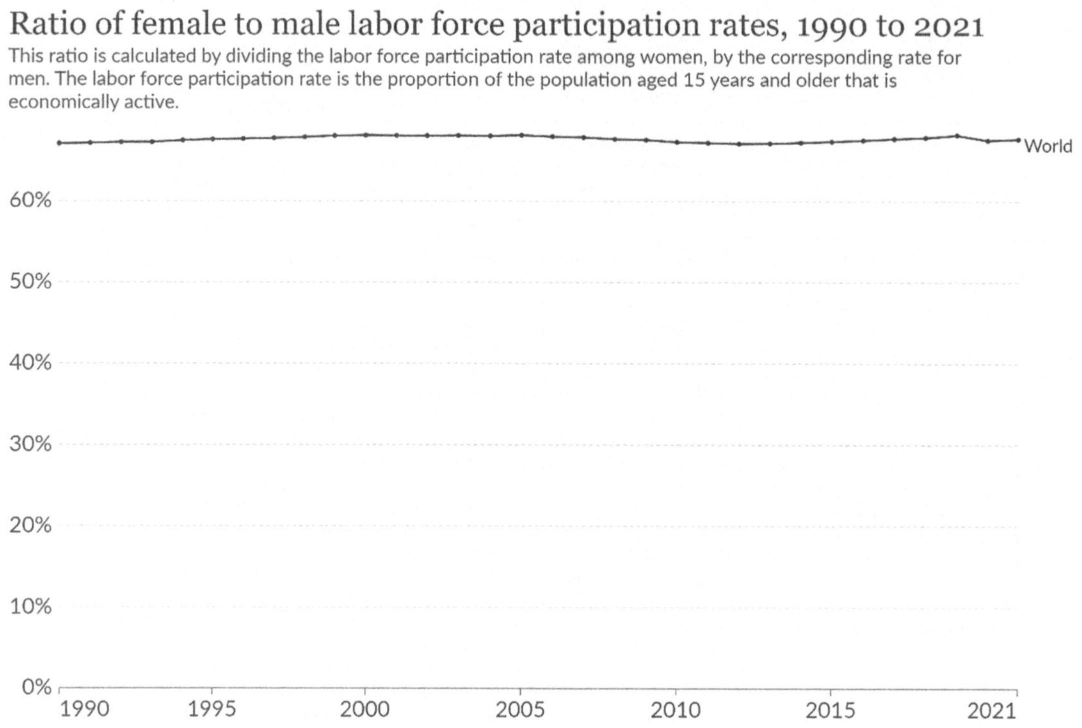

Fig. 2.18 Female versus male labor force participation globally

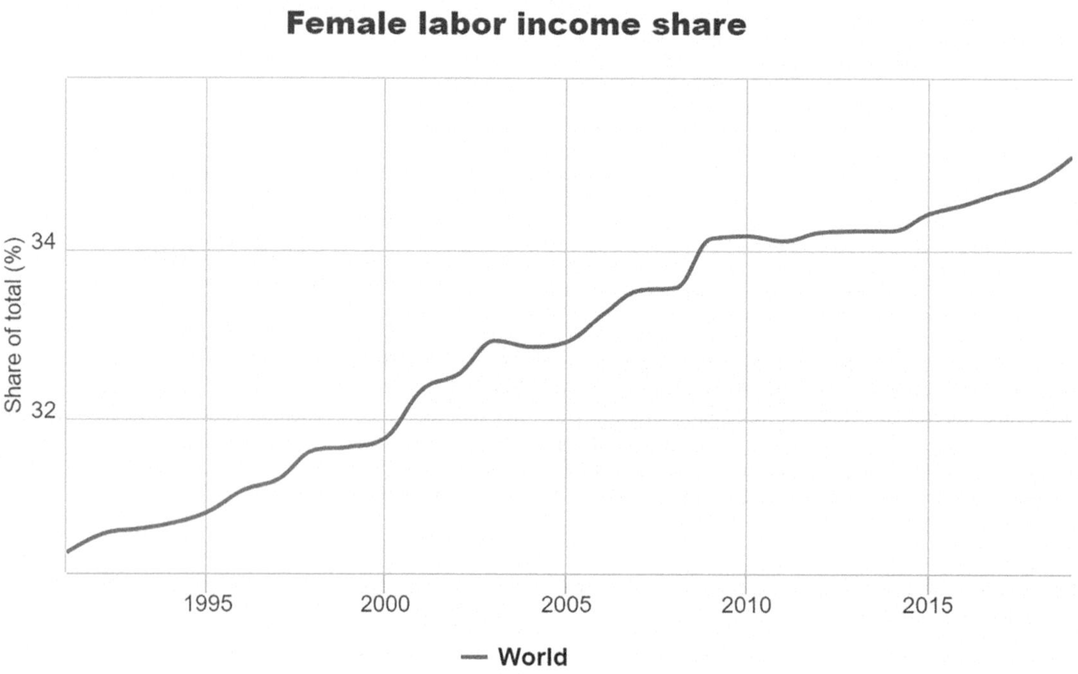

Fig. 2.19 Female labor income as share of total

substantial inequalities. These inequalities are a significant threat to the social pillar of sustainability. There are many other dimensions of social sustainability as well, such as race and caste, that apply to particular parts of the world, but gender inequality is of universal significance.

In the analysis of the views on sustainability of the business and political leaders of five countries—Guyana, Canada, India, Sri Lanka, and the USA—that are the focus of this study, indicators of economic, environmental, and social sustainability will be examined in providing a context for the views of the leaders.

References

Burford G, Hoover E, Velasco I, Janouskova S, Jimenez A, Piggot G, Harder MK (2013) Bringing the 'missing pillar' into sustainable development goals: Towards intersubjective values-based indicators. Sustainability 5(7):3035–3059

Chitiga, M., Sekyere, E., & Tsoanamatsie, N. (2014). Income inequality and limitations of the Gini index: the case of South Africa. Human Sciences Research Council, Pretoria. From http://www.hsrc.ac.za/en/review/hsrc-review-november-2014/limitations-of-gini-index. Accessed 24 July 2019

Davies J, Lluberas R, Shorrocks A (2022) Credit Suisse global wealth databook. Credit Suisse, Zurich

Diaz S, Demissew S, Carabias J, Joly C, Lonsdale M, Ash N et al (2015) The IPBES Conceptual Framework—connecting nature and people. Curr Opin Environ Sustain 14:1–16

Elkington J (2018) 25 years ago I coined the phrase 'triple bottom line'. Here's why it's time to rethink it. HBR, June 25. From https://hbr.org. Accessed 30 Jan 2019

Hallward-Driemer MT, Hasan AB, Rusu (2013) Women's legal rights over 50 years: Progress, stagnation or regression? World Bank Policy Research Working Paper No. 6616. The World Bank, Washington, DC

Haughton J, Khandker S (2009) The handbook of poverty and inequality. The World Bank, Washington, DC

Hindle T (2009) Triple bottom line – it consists of three Ps: profit, people and planet. The Economist, November 17. From http://www.economist.com/node/14301663. Accessed 12 Feb 2019

Kelley C, Mohtadi S, Cane M, Seager R, Kushnir Y (2015, March) Climate change and the recent Syrian drought. Proceedings of the National Academy of Sciences 112(11):3241–3246

Kendi I (2020) Stop blaming black people for dying of the coronavirus. The Atlantic, April 14. From https://www.theatlantic.com/ideas/archive/2020/04/race-and-blame/609946/. Accessed 22 Apr 2020

Klasen S (2020) From 'MeToo' to BokoHaram: a survey of levels and trends of gender inequality in the world. World Dev 128:Article 104862, 1–10

Luhby T (2019) The top 26 billionaires own $1.4 trillion—as much as 3.8 billion other people. CNN, January 21

Millennium Ecosystem Assessment (2005) Ecosystems and human well-being: a framework for assessment. Island Press, Washington, DC

New York Times (2019) A quarter of humanity faces looming water crises. August 6. From https://www.nytimes.com/interactive/2019/08/06/climate/world-water-stress.html?auth=linked-google. Accessed 4 Apr 2020

Our world in Data (2023) Where does our electricity come from. From https://ourworldindata.org/electricity-mix. Accessed 26 July 2023

Oxfam International (2023) Richest 1% bag nearly twice as much wealth as the rest of the world put together over the past two years. From https://www.oxfam.org/en/press-releases/richest-1-bag-nearly-twice-much-wealth-rest-world-put-together-over-past-two-years. Accessed 27 Aug 2023

Pazienza M, de Jong M, Schoenmaker D (2022) Clarifying the concept of corporate sustainability and providing convergence for its definition. Sustainability 14:7838. https://doi.org/10.3390/su14137838

Quinney M (2020) COVID-19 and nature are linked. So should be the recovery. World Economic Forum. From https://www.weforum.org/agenda/2020/04/covid-19-nature-deforestation-recovery/. Accessed 22 Apr 2020

Solt F. (2019) Measuring income inequality across countries and over time: the Standardized World Income Inequality Database. SWIID Version 8.1, May 2019. From https://osf.io/preprints/socarxiv/mwnje/. Accessed 12 June 2019

Sridhar K, Jones G (2013) The three fundamental criticisms of the triple bottom line approach: An empirical study to link sustainability reports in companies based in the Asia-Pacific region and TBL shortcomings. Asian J Bus Ethics 2(1):91–111

The Amazon is approaching an irreversible tipping point (2019) The Economist, August 3, pp 14–16

UNEP (2019) Global environment outlook-GEO-6: healthy planet, healthy people. Cambridge University Press, Cambridge

Winfield N (2020) Pope dreams of post-virus world where inequalities abolished. ABC News, April 19. From https://abcnews.go.com/Business/wireStory/pope-dreams-post-virus-world-inequalities-abolished-70230300. Accessed 22 Apr 2020

World Bank (2020) World development indicators database. From https://databank.worldbank.org/views/reports/reportwidget.aspx?Report_Name=CountryProfile&Id=b450fd57&tbar=y&dd=y&inf=n&zm=n&country=CAN. Accessed 10 July 2020

World Bank (2022) World development indicators database. World Bank, Washington, DC. From http://data.worldbank.org. Accessed 17 July 2023

World Inequality Database (2023). From https://wid.world/world/#sptinc_p90p100_z/WO/last/eu/k/p/

yearly/s/false/51.627/60/curve/false/country. Accessed 20 July 2023

WWF (2020) Water scarcity. From https://www.worldwildlife.org/threats/water-scarcity. Accessed 4 Apr 2020

Yeung J (2020) Australia's deadly wildfires are showing no signs of stopping. Here's what you need to know. CNN, January 13. From https://www.cnn.com/2020/01/01/australia/australia-fires-explainer-intl-hnk-scli/index.html. Accessed 1 Apr 2021

3 Sustainable Development Goals

Abstract

As 2030 draws closer there is increased interest in the United Nations Sustainability Development Goals. These 17 goals were adopted by member states of the UN in 2015 with the hope of achieving them by 2030. This chapter provides an overview of the 17 goals that are related to the pressing social, economic, and environmental issues together with an assessment of them being achieved by the deadline. At this point, it appears as if most of these minimum sustainability targets will not be met.

The United Nations Sustainable Development Goals were adopted by the UN at a special summit of 193 member states in September 2015. The 17 goals were meant to be achieved by 2030 and cover the general areas of the Environment, Economy, and Social issues. The Sustainability Development Goals are described in various publications by international agencies. The United Nations Development Programme is prominent among these, and it provides details on statistics and timelines for the achievement of the 17 interrelated goals (UNDP 2020).

A description of the UN's 17 Sustainability Development Goals (UN 2019) and progress in achieving them follows.

3.1 Goal 1: No Extreme Poverty by 2030

The goal of no poverty is to eradicate extreme poverty by 2030. Some progress has been made in achieving this goal, mainly due to economic growth in India and China.

"As of 2015, about 736 million people still lived on less than US$1.90 a day; many lack food, clean drinking water and sanitation" (UNDP 2020). The bulk of those still living in extreme poverty is in South Asia and sub-Saharan Africa. There has been a steady decline in poverty over the past three decades and from 2015 to 2018 the global poverty rate fell from 10.1% to 8.6% (UN 2023). However, progress was interrupted in the 2018–2020 period when the poverty rate rose to 9.2%, a setback attributed to the COVID-19 pandemic. The UN estimates that if current trends continue "575 million people will still be living in extreme poverty and only one-third of countries will have halved their national poverty levels by 2030" (UN 2023).

J. Singh, *Sustainability and the Sustainable Development Goals in Cultural Clusters*, Sustainable Development Goals Series, https://doi.org/10.1007/978-3-031-70618-9_3

3.2 Goal 2: Zero Hunger, Food Security and Sustainable Agriculture

According to the UNDP, as of 2017, there were approximately 821 million people in the world who were chronically undernourished. Although still shockingly high, the number of chronically undernourished people has dropped by almost half in the last two decades. The goal of no hunger is to eradicate hunger and malnutrition by 2030. However, it appears highly unlikely that this goal will be achieved. Many factors including the Russia/Ukraine war and weather conditions in India have contributed to a deteriorating food supply chain. Moreover, the impact of COVID-19 is estimated to be an additional 132 million chronically hungry human beings (UN News 2020).

3.3 Goal 3: Good Health and Well-Being

This goal aims to promote good health globally as a key ingredient of sustainable development. Although much progress has been made in the battle against illnesses—such as HIV and malaria—progress has not been uniform, and disparities exist within and between countries and regions. The advent of COVID-19 and the disparate response globally will certainly impede the achievement of this goal by 2030. Like the other goals, several targets were set for this goal. Some of these targets may be met by 2030 but not all. For example, the target for under-5 mortality will likely be met by most countries but that for maternal mortality likely will not.

3.4 Goal 4: Quality Education and Lifelong Learning

There has been remarkable progress in achieving universal primary education. The goal, with a target of 2030, is for all boys and girls worldwide to complete primary and secondary education. Between 2000 and 2015, enrolment in developing countries rose to 91%, and according to the UNDP, by 2015, there was a dramatic decrease in the number of children dropping out. There has been progress on this goal, but it is not likely that by 2030 all boys and girls worldwide will complete primary and secondary education. Worldwide, the primary school completion rate had risen to 85% in 2019 from 70% in 2000. Completion rates for lower secondary school was 73% and upper secondary school 49%. Students in wealthier countries were more likely to complete primary and secondary education, and within countries, this trend held (UN 2020). Goal 4 also seeks to provide universal access to vocational and higher education. Progress in achieving this goal has also been severely impeded by the COVID-19 pandemic. "It is estimated that 147 million children missed more than half of their in-class instruction over the past two years. This generation of children could lose a combined total of $17 trillion in lifetime earnings in present value" (UN 2023).

3.5 Goal 5: Gender Equality

The UNDP, together with other international agencies and governments worldwide, views gender equality as a basic human right. Goal 5 aims to end all discrimination against women and girls by 2030. Progress has been made in areas like education and labour market participation, but much remains to be done in some parts of the world. This goal too is unlikely to be achieved by 2030. With less than seven years remaining, a mere 15.4 percent of Goal 5 indicators with data are 'on track,' 61.5 percent are at a moderate distance and 23.1 percent are far or very far off track from 2030 targets (UN 2023).

3.6 Goal 6: Clean Water and Sanitation

It is difficult for many people in developed countries to imagine living with water scarcity, but according to the UNDP, this is the case for 40% of the world's population, and this figure is expected to increase with rising temperatures. The goal of safe and affordable drinking water for all by 2030 poses a challenge, but it is critical to sustainability. Also important is the provision of adequate sanitation services as, "in 2015, 4.5 billion people lacked safely managed sanitation services (with adequately disposed or treated excreta), and 2.3 billion lacked even basic sanitation" (UNDP 2020). Achievement of this goal by 2030 is in jeopardy. It appears as if a substantial segment of the world's population will face water shortages and a lack of proper sanitation for decades to come. The UN (2023) estimates that achievement of this goal by 2030 will require a sixfold increase in current rates of progress for "drinking water, fivefold for sanitation and threefold for hygiene."

3.7 Goal 7: Affordable and Clean Energy

Is the goal of affordable and clean energy achievable by 2030? With less than a decade to go, it doesn't appear likely. Almost 90% of the world's population has electricity, and since much of this is produced with fossil fuels, including large amounts of coal in India and China, it is a major cause of global warming. There has been some progress on renewable electricity sources but only 30% of the global electricity supply is currently produced this way. It will be a major challenge to move vast numbers of people away from using energy produced from dirty sources to using clean energy from renewable sources.

3.8 Goal 8: Decent Work and Economic Growth

Although the world's economy has recovered from the 2008 meltdown, the economic devastation resulting from the COVID-19 pandemic will severely impede the achievement of Goal 8 by 2030. Hundreds of millions of people worldwide have lost their jobs or are working reduced hours because of the pandemic. Before the pandemic, steady progress was being made in reducing extreme poverty and in creating jobs for men and women worldwide—although the participation rate of the latter continues to lag behind that of men. Stimulus initiatives in several countries aimed at economic recovery from the COVID-19 slowdown have led to an increase in the global inflation rate. The raising of interest rates to combat inflation has resulted in economic hardship for many.

3.9 Goal 9: Industry, Innovation, and Infrastructure

Goal 9 aims to promote industry, innovation, and infrastructure that is conducive to sustainable development. While there has been progress on this front, 2.6 billion people in developing countries do not have a reliable supply of electricity, and more than half of the world's population does not have access to the Internet. Progress on this goal is uneven with the least developed countries unlikely to meet the target of doubling the contribution of the manufacturing sector by 2030.

3.10 Goal 10: Reduced Inequality

"Income inequality is on the rise—the richest 10% have up to 40% of global income, whereas the poorest 10% earn only between 2% to 7%" (UNDP 2020). Addressing income inequality is important because it undermines the economic system by sucking resources out of circulation; is not sustainable; and is downright unfair. More data were earlier provided on the global inequality of income and wealth. It is a topic that receives constant scrutiny in the press and has moved politicians in Canada to at least talk about new approaches, such as a guaranteed minimum income. Whether this stop-gap measure will reach fruition in Canada is yet to be seen, but there is significant pressure on politicians to address the perceived failings of capitalism.

This goal seeks to reduce income and other inequalities and is clearly within the purview of business and political leaders. There has been some progress on this goal, but again the disruptions engendered by COVID-19 put the attainment of this goal at risk. One target included in this goal is the income of the bottom 40% of the population should grow faster than national averages. Preliminary estimates from the United Nations suggest that COVID has disrupted progress being made in achieving this target. Moreover, the inequality of income among nations has increased. This has contributed to the growing number of refugees worldwide.

3.11 Goal 11: Sustainable Cities and Communities

It is expected that by the year 2050, two-thirds of the world's population of close to 10 billion people will be in urban areas. The UNDP and other agencies recognize that changes in the way cities are designed will have to be implemented for the migration to cities to be sustainable. "Making cities sustainable means creating career and business opportunities, safe and affordable housing, and building resilient societies and economies. It involves investment in public transport, creating green public spaces, and improving urban planning and management in participatory and inclusive ways" (UNDP 2020).

"Goal 11 is about making cities and human settlements inclusive, safe, resilient and sustainable" (UN 2023). Many people in urban areas will live in cramped conditions and environmentally unhealthy conditions. COVID-19 will also affect the realization of Goal 11. The initial improvement in air pollution resulting from lockdowns will likely be reversed as economies reopen, and governments may even be tempted to relax environmental regulations to facilitate higher output. "In 2016, 9 in 10 people living in urban areas worldwide were breathing air that did not meet the World Health Organization air quality guidelines value for particulate matter (PM2.5). That same year more than half the world's urban population was exposed to air pollution levels at least 2.5 times above that value. Half the world's urban population also experienced an increase in particulate matter and a worsening of air quality from 2010 to 2016" (UN 2020). The 2023 forest fires in Canada, Greece, and elsewhere have likely caused further deterioration of global air quality, including in urban areas.

3.12 Goal 12: Responsible Consumption and Production

This Sustainable Development Goal aims to improve the way goods are produced and consumed. It is believed that in an interconnected world, shared resources must be managed in a way that does not favor one country or region over others. The disproportionate use of resources by the developed Western nations is therefore seen as being unsustainable. A more efficient approach to the production of goods and subsequent disposal of waste is one way of enhancing sustainability. Measurement of progress in reaching targets associated with this goal is not precise but, in many cases, simply ascertain whether a country has a particular program or policy. Awareness of the need for responsible consumption and production appears to be increasing. For example, most cities around the world have recycling programs, and while these are seen as a bare minimum effort in responsible waste management, it could be argued that it is an indication of increased awareness of the need for responsible production and consumption. However, there remain serious challenges in attaining this goal. This is exemplified in the world's approach to the use of fossil fuels, which are highly polluting and the top contributor to global warming. However, governments around the world continue to subsidize the use of fossil fuels. "Fossil fuel subsidies globally rose to over $400 billion in 2018, approaching 2014 levels, but fell by around 27% in 2019, mostly due to a decline in fuel prices" (UN 2020). However, the fallout from the COVID-19 pandemic and the Russia/Ukraine war caused a doubling of such subsidies from 2020 to 2021.

3.13 Goal 13: Climate Action

Global warming is a serious threat to the planet and left unchecked will make sustainable development impossible. The foregoing sections of this book provide details on the rise in greenhouse gas emissions and what needs to be done to reverse global warming. It cannot be overemphasized that fossil fuels are the main cause of global warming and their use must be urgently curbed. There is some indication that the world may be turning to renewable energy sources, but the use of fossil fuels continues to provoke debate on fairness in tackling what is very widely seen as a global issue. The Paris Agreement seeks to limit global warming to 1.5 °C above the preindustrial level, a target that is contingent on greenhouse gas emissions peaking before 2025 and decreasing to net zero by 2050. One issue that has provoked much debate is the role of developed countries in fighting global warming. Developing countries have argued that developed countries should contribute more to fighting global warming as they have contributed disproportionately to the cause of the problem. The developed countries generally agree with this position and have pledged to raise US$100 billion per year by 2020 through 2025 for a Green Climate Fund. However, they have fallen short of this commitment as shown in Fig. 3.1.

3.14 Goal 14: Life Below Water

Goal 14 is aimed at making the world's oceans healthy again. Billions of people around the world are reliant on oceans and coasts for their livelihood. Moreover, oceans are carbon sinks as they absorb carbon dioxide produced by human beings. About a quarter of the planet's carbon emissions is absorbed by oceans, and this has led to a 26% drop in their pH level since preindustrial times (UN 2020). Therefore, this process helps to mitigate global warming but is destructive to life below water. The oceans are at a breaking point, and restoring their health is an integral part of the sustainability development agenda. According to

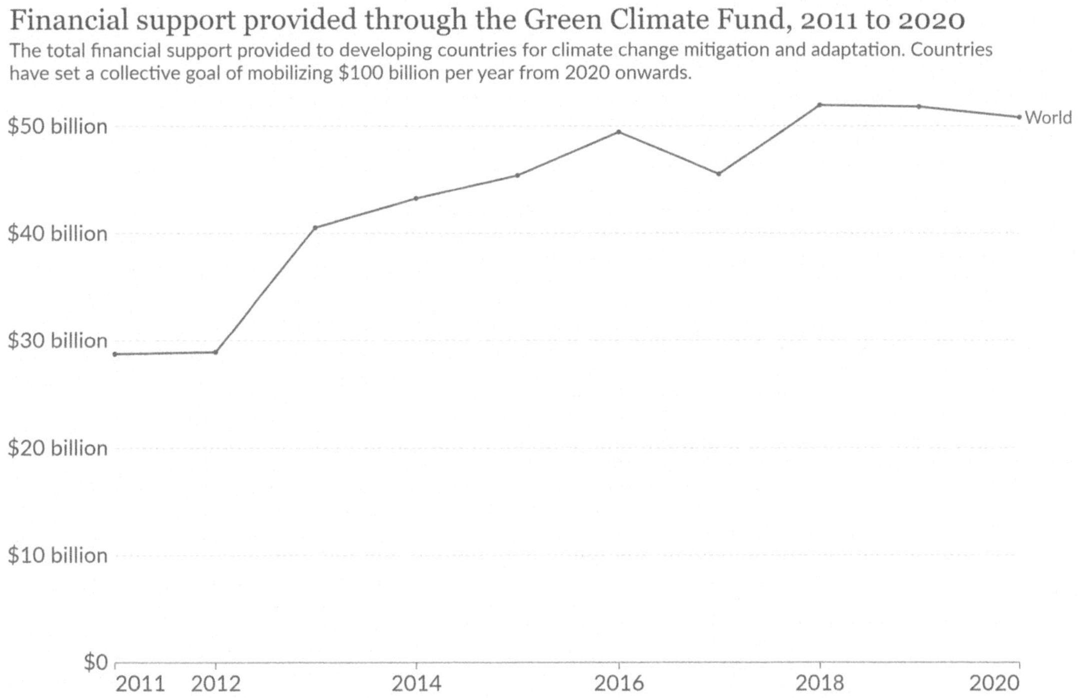

Fig. 3.1 Green climate fund (Our World in Data 2023)

the UN, progress towards the attainment of Goal 14 is not on track to meet targets by 2030.

3.15 Goal 15: Life on Land

"Goal 15 is about conserving life on land. It is to protect and restore terrestrial ecosystems, sustainably manage forests, combat desertification, halt and reverse land degradation, and stop biodiversity loss" (UN 2023). Encouraged by the irresponsible words and actions of leaders like former Presidents Jair Bolsonaro of Brazil and Donald Trump of the USA, businesses and individuals have destroyed some of the world's most precious diversity and upset the balance of nature on land. Foremost among these is the Amazon rainforest, the bulk of which is located in Brazil. However, the degradation of biodiversity on land is not limited to the Amazon rainforest. It is happening throughout the world, and arresting this decline is critical to sustainability.

3.16 Goal 16: Peace, Justice, and Strong Institutions

"We cannot hope for sustainable development without peace, stability, human rights, and effective governance, based on the rule of law" (UNDP 2020). This and the other Sustainable Development Goals aim to achieve peace and stability in all nations, as sustainable development is not possible in an unstable environment. While some nations and regions seem to be constantly embroiled in conflicts, international organizations like the UNDP believe that peace, stability, and respect for the rule of law are achievable by all nations. However, recent developments do not bode well for this aspiration. The continuing Russia/Ukraine war has created havoc in the lives of people in those countries and elsewhere. Military coups in Gabon and Niger have compromised the rule of law in those countries and military conflicts in Sudan and Gaza have displaced hundreds of thousands. Urgent international action to resolve these and other conflicts is imperative if there is to be any hope of achieving Goal 16 by 2030.

3.17 Goal 17: Partnerships for the Goals

Remarkably, virtually all the countries of the world agreed to the Sustainable Development Goals. The level of cooperation that led to agreement on the goals must be maintained for the achievement of the goals by 2030. However, crises like that engendered by the COVID-19 pandemic and the Russia/Ukraine war are not predictable and will severely test continuing cooperation. As clearly illustrated by the pandemic, only a coordinated, cooperative global approach can effectively deal with global challenges in an interconnected world.

Many of the SDGs are aimed at reducing social and economic inequality, and some are aimed at ensuring the survival of the planet itself. In this book, for each of the five countries from which interviewees were drawn, the state of social and economic inequality will be examined through various widely used indices. The state of the physical environment in these countries will also be examined. The objective of examining these measures is to position Guyana, Canada, India, Sri Lanka, and the USA on sustainability indices to place the opinions of the interviewees within the context of their countries' performance on the three pillars of sustainability.

References

Our World in Data (2023) SDG tracker: measuring progress towards the Sustainable Development Goals. Published online at OurWorldInData.org. From https://ourworldindata.org/sdgs. Accessed 29 Aug 2023

UN (2019) Sustainability Development Goals. From https://www.un.org/sustainabledevelopment/. Accessed 11 Feb 2019

UN (2020) The Sustainable Development Goals report. From https://unstats.un.org/SDG/report/2020. Accessed 18 Nov 2020

UN (2023) Sustainability Development Goals. From https://www.un.org/sustainabledevelopment/sustainable-development-goals/Accessed 1 Sept 2023

UN News (2020) UN report sends 'sobering message' of deeply entrenched hunger globally. From https://news.un.org/en/story/2020/07/1068261. Accessed 19 Nov 2020

UNDP (2020). Sustainability Development Goals. From https://www.undp.org/content/undp/en/home/sustainable-development-goals. Accessed 3 Oct 2020

Case Study 1: United States of America

4

Abstract

This is the first of five chapters that analyze the interviews conducted with business and political leaders in selected countries. In this chapter interviews with leaders in the USA are examined. The chapter begins with background information on the country's history, economy, people, and environment. This is followed by the identification and discussion of themes on sustainability that permeate the interviews. The themes identified include long-term profitability, educating the people, adapting to change, employment of women, income inequity among races, lack of political leadership, opportunity gap, experiences as a developed country, all countries taking part, and consequences of actions.

The United States of America (USA) is facing monumental problems that are directly or indirectly linked to sustainability. The country born in rebellion against British colonialism and built on the ideals of life, liberty, and the pursuit of happiness in 2024 is facing potentially existential threats. On the economy, the GDP of the world's largest economy had shrunk by unprecedented proportions, and at the end of July 2020, the country was officially in an economic recession. It pulled out of the recession but experienced such high inflation that the Federal Reserve responded with high interest rates. Unemployment had risen sharply during the pandemic and some economists were warning of people going hungry and without shelter. However, at the same time, the very rich have been getting richer. For example, in just 1 day in July 2020, Jeff Bezos' wealth increased by US$7 billion.

The dismal economic conditions are partly due to COVID-19 as businesses were forced to close or curtail their operations. At the end of July 2020, the USA with just 4% of the world's population, accounted for about 25% of the world's COVID-related deaths. Moreover, the victims of COVID have been disproportionately Black and Brown, a reality many ascribe to living conditions attributable to a lopsided economic system that has created severe inequities and an inequitable and disjointed health care system.

Moreover, there were sustained protests in several large US cities, demanding racial justice and police reform. These protests were sparked by the murder by Minneapolis police officers of a Black man George Floyd on May 25, 2020. Amid the economic and social challenges, an insensitive and divisive President Donald Trump desperately sought to overturn the 2020 election that resulted in his defeat by President Joe Biden. Reverberations from the January 6, 2021, storming of Capitol Hill are still being felt and Donald Trump and several associates are on trial for allegations of criminal behavior in their attempt to

J. Singh, *Sustainability and the Sustainable Development Goals in Cultural Clusters*, Sustainable Development Goals Series, https://doi.org/10.1007/978-3-031-70618-9_4

prevent Joe Biden's ascension to the highest office in the USA. As the world's largest economy and a military superpower, what happens in the USA is of great concern to the rest of the world.

The USA came into being in 1776, when 13 North American colonies rebelled against British rule to form a country that was subsequently joined by 37 other states. Together the 50 states (and the District of Columbia) cover 9,833,517 km^2. It is a vast nation with a diverse topography: large urban centers, extensive national parks, and substantial agricultural areas.

The population of the USA is estimated to be 339,665,118 in 2023 (CIA 2023). While the USA has been described as a melting pot of cultures, in reality, it is a multicultural society according to various dimensions. The racial composition of the USA population in 2010 was: White 72.4%; Black 12.6%; Asian 4.8%; Amerindian and Alaska Native 0.9%; Native Hawaiian and other Pacific islander 0.2%; other 6.2%; two or more races 2.9% (CIA 2020).

While most Americans list English as the language they speak at home, there is diversity in the first language of Americans as indicated by the following figures: English only 78.2%; Spanish 13.4%; Chinese 1.1%; other 7.3% (2017 est.), (CIA 2020). Further diversity is found in the religious affiliation of Americans as follows: Protestant 46.5%; Roman Catholic 20.8%; Jewish 1.9%; Mormon 1.6%; other Christian 0.9%; Muslim 0.9%; Jehovah's Witness 0.8%; Buddhist 0.7%; Hindu 0.7%; other 1.8%; unaffiliated 22.8%; don't know/refused 0.6% (2014 est.) (CIA 2020).

The USA is governed by a federal government, 50 state governments, and thousands of municipal governments in a system where the levels of government have jurisdiction in defined areas. The USA is a very wealthy nation with a GDP of approximately US$21.1 trillion and a per capita GDP of US$63,700 in 2021 (CIA 2023). However, these numbers do not tell the story of income and wealth inequality. This and figures on other aspects of sustainability are shown in the USA Country Profile that follows in Table 4.1. On a composite score of progress in achieving the 17 SDGs the USA scores 75.91 where 100 would indicate full achievement of all the goals (Sachs et al. 2023). The country ranks 39th among 166 countries.

The perceptions of American leaders on sustainability are presented in this section. The themes that emerged from the analysis are presented in Table 4.2. The themes are further described in the following subsections.

4.1 Narrative Themes from USA Leaders

4.1.1 Long-Term Profitability

American business leaders interviewed believed that economic sustainability involved long-term profitability. For long-term profits, leaders needed to be strategic in spending in addition to saving money. Participant UB1 commented:

> *On the other hand, I think business has taken up the role, and I don't believe they're turning back. I think that they have found profitability, sustainability. The example you used of solar panels on warehouses and in selling the electricity back over a long period of time has given new businesses a way to be profitable, and it's also given the recipient of that warehouse electricity, cheaper electricity. So, if you look at this city now, for a long time, we were the number one solar city in the United States, and I think that was because the city is culturally very aware of the environment because of our beaches and just where we sit in the United States. And I think that, that's given people and business an opportunity to learn how to be profitable by using sustainable practices. I don't see anybody turning back. I see it accelerating to a great extent, and I think that ties into the culture because people expect businesses to be responsible, to be sustainable too by end of the sustainability issue, and I think that it just starts circling.*

Participant UB3 claimed that sustainability involved being able to afford changes. The participant stated:

> *Change costs profits. On the other hand, electric cars are a new profit centre for us. It's a good thing. We've had to change, to survive, and if we don't ... From a political standpoint, we've had to change. Richard Nixon said, we need to become oil independent in the United States. We laughed at him, and we never thought of it. It's impossible, and we can't do it. Today, we are 100% in the United States oil independent; we're way down the road.*

Table 4.1 Country profile: the USA

Country profile				
	1990	2000	2010	2021
World view				
Population, total (millions)	249.62	282.16	309.33	332.03
Population growth (annual %)	1.1	1.1	0.8	0.2
Surface area (km^2) (thousands)	9629.10	9632.00	9831.50	9831.50
Population density (people per km^2 of land area)	27.3	30.8	33.8	36.2
Poverty headcount ratio at national poverty lines (% of population)	–	–	–	–
Poverty headcount ratio at $2.15 a day (2017 PPP) (% of population)	0.5	0.7	1	0.2
GNI, Atlas method (current US$) (billions)	6006.22	10,146.58	15,202.01	23,539.92
GNI per capita, Atlas method (current US$)	24,060	35,960	49,150	70,900
GNI, PPP (current international $) (billions)	5902.29	10,383.67	15,172.07	23,617.11
GNI per capita, PPP (current international $)	23,640	36,800	49,050	71,130
People				
Income share held by lowest 20%	5.6	5.5	5.3	5.6
Life expectancy at birth, total (years)	75	77	79	76
Fertility rate, total (births per woman)	2.1	2.1	1.9	1.7
Adolescent fertility rate (births per 1000 women ages 15–19)	61	47	34	16
Contraceptive prevalence, any method (% of married women ages 15–49)	71	73	76	74
Births attended by skilled health staff (% of total)	99	99	99	99
Mortality rate, under-5 (per 1000 live births)	11	8	7	6
Prevalence of underweight, weight for age (% of children under 5)	0.8	1.1	1	0.4
Immunization, measles (% of children ages 12–23 months)	90	91	92	91
Primary completion rate, total (% of relevant age group)	–	–	–	101
School enrollment, primary (% gross)	106	101.9	99.5	100.3
School enrollment, secondary (% gross)	91	94	96	101
School enrollment, primary and secondary (gross), gender parity index (GPI)	1	1	1	1
Prevalence of HIV, total (% of population ages 15–49)	–	–	0.4	0.4
Environment				
Forest area (km^2) (thousands)	3024.50	3035.40	3087.20	3098.00

(continued)

Table 4.1 (continued)

Country profile				
	1990	2000	2010	2021
Terrestrial and marine protected areas (% of total territorial area)	–	–	–	15.9
Annual freshwater withdrawals, total (% of internal resources)	16.5	16.8	14.9	15.8
Urban population growth (annual %)	1.4	1.5	1	0.4
Energy use (kg of oil equivalent per capita)	7672	8057	7161	–
CO_2 emissions (metric tons per capita)	19.41	20.47	17.43	13.03
Electric power consumption (kWh per capita)	11,713	13,671	13,395	–
Economy				
GDP (current US$) (billions)	5963.14	10,250.95	15,048.96	23,315.08
GDP growth (annual %)	1.9	4.1	2.7	5.9
Inflation, GDP deflator (annual %)	3.7	2.3	1.2	4.5
Agriculture, forestry, and fishing, value added (% of GDP)	–	1	1	1
Industry (including construction), value added (% of GDP)	–	22	19	18
Exports of goods and services (% of GDP)	9	11	12	11
Imports of goods and services (% of GDP)	11	14	16	15
Gross capital formation (% of GDP)	22	24	19	21
Revenue, excluding grants (% of GDP)	18.2	20.6	16.4	18.7
Net lending (+)/net borrowing (−) (% of GDP)	−3.7	2.2	−10	−12.8
States and markets				
Time required to start a business (days)	–	–	6	4
Domestic credit provided by financial sector (% of GDP)	145.6	191.6	218	289.3
Tax revenue (% of GDP)	10.5	13	8.6	11.2
Military expenditure (% of GDP)	5.6	3.1	4.9	3.5
Mobile cellular subscriptions (per 100 people)	2.1	38.8	91.6	107.3
Individuals using the internet (% of population)	0.8	43.1	71.7	91.8
High-technology exports (% of manufactured exports)	–	–	23	20
Statistical capacity score (overall average) (scale 0–100)	–	–	–	–
Global links				
Merchandise trade (% of GDP)	15	20	22	20
Net barter terms of trade index (2000 = 100)	103	100	97	100
External debt stocks, total (DOD, current US$) (millions)	–	–	–	–

Total debt service (% of exports of goods, services and primary income)	–	–	–	–
Net migration (thousands)	1422	1480	1031	562
Personal remittances, received (current US$) (millions)	1170	4395	5930	6899
Foreign direct investment, net inflows (BoP, current US$) (millions)	71,230	349,125	264,039	448,324
Net official development assistance received (current US$) (millions)	–	–	–	–

World Bank (2023)
Source: World Development Indicators database
Country: United States Data from database: WDI Central

Table 4.2 Major themes from USA

Pillar of sustainability	Business leaders	Political leaders
Economic	Long-term profitability	Educating the people
	Educating the people	Adapting to change
Social	Employment of women	Opportunity gap
	Income inequity among races	Employment of women
	Lack of political leadership	
Environmental	Experiences as a developed country	All countries taking part
		Consequences of actions

On the theme of long-term profitability, participant UB1 also addressed the issue of energy self-sufficiency:

> *I think that it comes down to a couple of things. It comes down to [a] number, dollars and cents, and if you can be 100% sustainable, that saves a tremendous amount of money in terms of infrastructure, in terms of importing energy, and in all of those ways. So I think that it's not necessarily like ideological. I think it's more people are expecting it. They're finding that it makes tremendous amount of sense, and they're learning how to make a profit from it.*

4.1.2 Educating the People

Business leaders UB1 and UB3 believed that providing education to the people increased economic sustainability, as more people earned income. Participant UB1 claimed:

> *In the United States, labour frequently runs K through 12 education, and labour caters to the teachers. It does not cater to the children. And then you have parents on the other side who want to see their children move rapidly, and usually those are the ones from better families, so they get it no matter what, and that leaves a huge gap in there for the kids who don't have a supporting parent or who can't.*

Participant UB4 had similar views, and stated:

> *What I'm trying to say is that with the economic freedom ... First, it's the education, which is more and more parents are wanting their kids and their child to go through the best education like their boys, that leads to them good jobs, they are economically free, they can make their own decisions, and they want their opinion to be counted. It does give rise to a good amount of, I would say, positive attention, because you just cannot ignore, or you cannot just say it doesn't matter. You go to the kitchen and just make the next meal.*

Political leaders also believed that educating people contributed to economic sustainability. UP1 believed that economic sustainability was job-based and that corporations could make changes to accommodate the wages in comparison to the cost of living. While the government had programs to support education, success would be dependent on additional factors. The participant stated:

> *Not necessarily [government policy]. I think it's also part of making sure that we have people who are going to be equipped to do the work. I sit on the governor's workforce development board. The thing is we make available the different programs, but you also have to have a willing public to engage. So there are many arms to all of this. There's the psychological. If you grew up in an environment where you believe that you did not have to do that because government will take care of you, then you won't have the desire or the motivation to be equipped to get a job. So it's many tentacles, but at the same time, there has to be the desire to do that. But, yes, we also need to make sure that we have the programs.*

Participant UP2 also addressed the link between education and economic sustainability. As he explained:

> *Oh, I think that one, simply put, no, it's not sustainable, and I would take it one further step. That wealth gap leads to an education gap. And I would argue certainly in our state the number one issue facing us is the education gap between the wealthy communities and the impoverished cities—that we're not getting the same education. And until we solve that, we're creating a bigger and bigger problem.*

Now, some of that is either caused by, or fixing that would lead to the wealth gap being closed. Because I could take the city right next door—that has a high unemployment rate, has a number of other economic issues, but consequently has a low cost of land and property—and I could argue that if we can fix the education gap, then everything else will take care of. You can pave roads for a fixed amount of money. You can solve crime for a fixed amount of money. It's the education gap. If you solve that, that will let the people there actually gain an equity because they're already owning the homes. Every time the value goes up, the incremental dollar is an incremental dollar of equity for the homeowner. That's the power of growing property values, in this case.

4.1.3 Adapting to Change

American political leaders emphasized the importance of adapting to change in the sustainability arena. The participants generally believed that the government should be flexible in making legislation for economic growth. Participant UP2 claimed:

We certainly are concerned when we read at the national level that the federal government is releasing some of the controls on coal because, as an old business associate of mine once said, "You used to fire up the coal plants, and you can count the number of people that die downwind." I mean, it's a very real number. And I'm an engineer at heart and by training, so that anything that's quantifiable gets my attention very fast. So, when we can save money and help the environment and create jobs, to me that's a three-way win for everybody, and I keep coming back to that because as I look at sustainability, our goal is not only to accomplish the goals and ... but it's also, then, turn around and then literally to market our town. Market this town as a sustainable community so that people that have that value set will move here, and that will reinforce what we're doing from a cultural standpoint.

The role of government in conserving natural resources was deemed important. Participant UP4 shared:

I don't think that there is enough emphasis given to sustainability, to renewability. I think that we know that we're dealing with finite limits with respect to arable land, or agricultural purposes, or a number of resources, a number of chemicals, a number of minerals, a number of metals, and a number of fossil fuel sources. We seem to recognize the fact that these are finite resources, but we are not doing enough to try and find a replacement or a supplement that will sustain the generations to come when we run out of those resources, whatever they may be.

4.1.4 Employment of Women

Both groups of American leaders perceived that women have experienced less employment inequity in recent years, but inequity remains a festering issue. On the positive side, more women are enrolled in higher education and are pursuing professional careers. Participant UP3 related a personal experience involving his daughter pursuing a career in law. The participant claimed that in terms of career opportunities, men and women were equal. However, in terms of income, women were still believed to experience lower pay. Participant UP4 shared:

If a woman applied for a job, I think there is a presumption that she is not entitled to the same as a man. It's been a cultural thing for as long as the United States has been in existence, and it's one of those things that we have to overcome. I was involved back in the 1980s when fair and equitable compensation to men and women became a very hot topic, where equity in compensation for both genders was seriously considered. I know that efforts have been made to try and create a greater equity, but we're far from being successful up till now in that regard.

From the perspective of the business leaders interviewed, women were generally doing well. According to them, many women were highly successful in their jobs and some women held leadership positions. Participant UB3 narrated:

We've certainly made a lot of progress, example: I go into the automotive business, 26 now... I've been out 11, so 37 years ago. There were no women. I come out of the travel business, which is 80% of women, and I find out that 80% of the bottom line decisions are made by women. Automotive dealers have a horrible reputation, guys in the automotive business are womanizers, so how are you going to get all them in? I went to an almost

> *all-women sales force and then, I let some men in. It was primarily women. My top salesperson last year was a woman, makes over $200 thousand a year. She's still among the top salespeople in my company.*

For participant UB2, the nature of the industry influenced the level of dominance of women and their opportunities for career advancement. The participant shared:

> *For instance, the health care industry tends to have more female nurses than males. So, I would say in my personal experience the women in the workforce, I would say there is pretty good equality in healthcare even up to the higher leadership areas. I mean, my executive team, over half the executive team are women today.*

4.1.5 Income Inequity Among Races

Acknowledging the history of racism in the United States, business leaders generally believed that minority groups still experienced income inequity. Participant UB2 claimed that minorities were not given opportunities for higher income, which increased the wage gap. The participant shared:

> *This is probably more of an academic, and certainly something you're exposed to maybe more than I. And it's certainly, as you said been raised more and more, particularly currently and probably because it's becoming a bigger and bigger disparity. So my gut would say probably this level of disparity probably isn't and needs to be adjusted although at least from what I see, the ones that have in theory, to moderate them or adjust them to actually solve the problem on the lower end, there's not enough money there.*

Participant UB3 explained that race also played a role in education, which then influenced income opportunities. The cycle involved preconceived notions on the performance of minorities. The participant discussed:

> *That has a lot to do with the African-American problem. I think, Hispanics have a similar issue in the United States. If your peers, your mothers, grandmothers, grandfathers have low wages and you're raised in low wages, it's a vicious circle at that point, versus you were brought up to be overachievers. To get into the University of California at San Diego as a non-Asian is very difficult because their genes aren't different, it's just they're brought up to be good students and study hard. When they get out of college, they're successful. If the African Americans, the Hispanics don't get to go to college; 60% of the kids in San Diego school district will not go to college.*

4.1.6 Lack of Political Leadership

The American business leaders interviewed shared the belief that the federal administration of that period had an impact on social sustainability. The lack of political leadership affected the society, and participants UB1 and UB4 believed that the administration caused a divide. Participant UB1 shared:

> *I guess there's two parts to that, politically speaking, no, because of Trump and what he said and a lot of people didn't know which way to believe about that because you believe what you want to believe. I think he has given a lot of people reason not to worry about things like that, so I think we've lost our leadership role, and I think that leadership role was critically important. As you said, there are other developing countries that if they see the United States has not committed to it, then they're going to think, "Why should we be committed to it?"*

4.1.7 Opportunity Gap

Political leaders also believed that minorities experienced opportunity gaps leading to employment and income inequity. Most of the participants explicitly stated that the opportunity gap was not socially sustainable. Participant UP2 highlighted:

> *So, I think that yes, there's a wealth gap, but I think it's an education that is the real issue, and that's what we have to solve. And you can't have a sustainable culture unless everybody's being educated so they can replenish the job market and be ready for the new jobs that come up. I was just at our teacher orientation today, and one of the short videos was by Google, that talked about all the things that needed to be done and might be invented in the*

future. And then they had pictures of young schoolchildren as being the ones that are going to invent the next time machine, if you will. That's what's going to happen there. So, I think that, to me, becomes one of the issues.

Participant UP2 stated that discrimination against minorities has existed for a long time in the USA and that technology has a role in mitigating racism. The participant explained:

One, it's not socially sustainable, what we're doing. That's where a number of these things are interdependent and come together. I could argue in some ways we should be more diverse and more equal than we've ever been. I'm not saying we are, by the way, but I think we could be because of all the education, all the freedom of movement. You're not stuck living in a town you grew up in. There are jobs. The Internet should be a great equalizer in terms of both letting you learn and perform without the bias of a human supervisor or evaluator.

Participant UP3, a legislator at the State level, was forthright in describing the opportunity gap and the vicious cycle of poverty it creates:

It's not the income gap. It's the opportunity gap that causes the income gap. And the reason I had this out is I was talking about the health access gap also. There's many gaps that cause a gap. The health access gap, if you look at who has dietary issues, it's generally poor people. Who is generally a poor person? It shows that it's a lot of times people who are Black. And what happens when they have bad diets? They get diabetes. And diabetes is very disabling and, many times, they will get so disabled that they have to go into long-term care. And in America, to go into long-term care, you have to be very wealthy, or you have to go on public assistance. And to go on public assistance, you need to spend down your wealth to a certain level. What that means is that when you die, you have nothing to give your heirs.

4.1.8 Experiences as a Developed Country

Business leaders believed that as a developed country, the USA had more access to regulations, solutions, and resources such as renewable energy. Environmental sustainability involves investing in sustainable materials, which may be a difficulty for developing economies. Participant UB1 shared:

Well, that's true, but I know that India is also exploring a new generation of nuclear technology which is much safer. It's much different; it's smaller units; it's completely contained. I know that Canada is exploring it, northern part of the country. So I think while India may say that in some of the most congested areas, I think they're also very interested in that. Maybe there has to be some type of fund to help with that to get it initially going. But they're going to have to do it, and I think that once that starts, that's just what takes off, and India, the number of engineers in India, it must be astounding.

Participants UB2 and UB3 believed that developed countries could help developing countries attain environmentally sustainable development. Participant UB3 stated:

Like I said, continue to do the things we're doing today, into the future and keep the planet intact.

Participant UB2 added:

But when I was using an example, it may not have been in India, but there are others who are doing better with water and you saying share technology. I think we all have to learn from each other. And I think to your point, we are all getting pushed closer to the precipice and that, actually, as with anything, and particularly if you relate to sustainability as with anything, that's painful or moving you in a wrong direction. You get pushed far enough, hopefully, you get smart enough to adjust.

4.1.9 All Countries Taking Part

Political leaders generally believed that all countries had environmental obligations. While developed countries could provide financial needs for sustainable development, Participant UP1 suggested that developing countries could contribute other resources. The participant stated:

Well, that's what I'm thinking. I mean there are so many wonderful scientists that come out of India. To say that India doesn't have the brain trust to do that is ridiculous ... is absolutely ridiculous. That's just, for me, again, my frustration with those kinds of negotiations. If we all come into all of this with the intention that we care about this planet, I don't care where we are, the moment is now, not 150 years in the past, not 150 years to the future. The moment is now for us to all say we are accountable.

Participant UP4 claimed that all countries could reduce waste, while participant UP2 stated that alternative energy could be introduced. Participant UP2 added that developing countries could learn from the mistakes of other countries. The participant narrated:

But you go back and look at our history, go to Pittsburgh. That was one of the worst cities and worst water contaminants ever, and they've cleaned that up. Long Island Sound, right out there, we're seeing more fish, different fish. We're seeing whales come in chasing those fish that we've never seen. I grew up here. We never had that. So, we're looking at all of that, and it's different. So, I think there are things we're making improvements on, but things we're ... If we haven't learned our lesson, we're repeating the same mistakes we made in this country.

The participant narrated that lessons could be learned from mistakes of the past, and this is on a global scale:

So, they had to clean up all of that, and they survived. Well, Beijing could've watched what happened to other cultures and their industrial revolution and not repeated the same mistakes, but they're not. When we went through our industrial revolution, and we had it here in town. We had a big battery factory. They made car batteries. One of the components of those batteries is lead. What they did with the excess lead is dump it in the river. So, we just spent, not the town but the company, millions of dollars, and over five years cleaned up the lead on the land, around the factory, and in the water. Well, if we hadn't dumped it there, we wouldn't have learned that. But if somebody else is now making those batteries and dumping the lead in the land and the water; they didn't learn from that. At some point you have to clean it up again. It's not sustainable because it gets into the food chain and sooner or later it hurts people.

4.1.10 Consequences of Actions

Political leaders believed that the current generation of leaders needed to be accountable for their actions and to think of the consequences for future generations. Participant UP1 shared:

Well, what that says to me is that so you took 150 years to do that, and the information we have is present, and the calamity that we are faced is present. Why do you say, "Well, it'll take us 150 years." Now, is there common sense in that kind of argument? For me, that's the kind of stuff that says then you really don't care. You really don't care. You're gonna say, "I don't have to address it. I'll let another generation address it."

Participant UP4 claimed that careless utilization of resources was not sustainable and that humans were relying on finite amounts of resources. The participant shared:

The simple answer is no, we can't continue the way we are, the current way. The best example is fossil fuels. We know that if it took billions of years to develop oil and natural gas and other fossil fuels, and we don't have billions of years to wait around for the reproduction of those fuels, we know that we have a finite limit to those fuels' availability.

4.1.11 Summary of Themes from Interviews with USA Leaders

The themes that emerged from the interviews with American political and business leaders were:

1. Long-term profitability
2. Educating the people
3. Adapting to change
4. Employment of women
5. Income inequity among races
6. Lack of political leadership
7. Opportunity gap
8. Experiences as a developed country
9. All countries taking part
10. Consequences of actions.

Business leaders in the United States believed that profitability sustained the economy. Both political and business leaders believed that educating the people helped provide employment opportunities to strengthen the economy. The political leaders also generally believed that changes needed to be experienced for the economy to be sustainable. In terms of social sustainability, employment and income inequity was believed to be experienced by minorities, mainly due to the lack of opportunities available to them. Women were experiencing less employment inequity and were starting to dominate and excel in certain fields. However, in terms of income, men still tend to be paid more than women. The business leaders perceived that the lack of political leadership was the source of the inequity and was causing a divide among the people.

On environmental sustainability, the business leaders perceived that developed countries could help developing countries, while political leaders believed that developing countries could play their part. Developed countries had the capacity to invest in sustainable materials while developing countries might contribute other resources, such as human resources. Political leaders believed that all leaders should consider the consequences of their actions and be accountable.

References

CIA (2020) The world factbook: USA. From https://www.cia.gov/library/publications/the-world-factbook/geos/in.html. Accessed 30 July 2020

CIA (2023) The world factbook: USA. From https://www.cia.gov/the-world-factbook/countries/united-states/. Accessed 6 Sept 2023

Sachs JD, Lafortune G, Fuller G, Drumm E (2023) Implementing the SDG stimulus. Sustainable development report 2023. SDSN/Dublin University Press, Paris/Dublin. https://doi.org/10.25546/102924

World Bank (2023) Country profile United States. From https://databank.worldbank.org/views/reports/reportwidget.aspx?Report_Name=CountryProfile&Id=b450fd57&tbar=y&dd=y&inf=n&zm=n&country=USA. Accessed 31 July 2023

Case Study 2: Canada

5

Abstract

This is the second of five chapters that analyze the interviews conducted with business and political leaders in selected countries. In this chapter interviews with leaders in Canada are examined. The chapter begins with background information on the country's history, economy, people, and environment. This is followed by the identification and discussion of themes on sustainability that permeate the interviews. The themes identified were closing the wealth gap, making and adapting to changes, employment equity and inclusion, effort of the private sector, and government action.

Canada is the world's second largest country by area, covering just below 10 million square kilometers. It is relatively sparsely populated with approximately 38 million inhabitants. The Confederation of Canada came into being in 1867 through a series of conferences. These meetings resulted in the former British colonies of New Brunswick, Nova Scotia, and Canada (later to become the provinces of Ontario and Quebec) establishing a confederation through the British North America Act that was approved by the United Kingdom parliament. The dominion was self-governed but with ties to Britain. It wasn't until 1982 that Canada's constitution was repatriated from the United Kingdom. Today Canada is a confederation of the provinces of Alberta, British Columbia, Manitoba, New Brunswick, Newfoundland and Labrador, Nova Scotia, Ontario, Prince Edward Island, Quebec, and Saskatchewan, and the territories of Northwest Territories, Nunavut, and Yukon.

At its founding in 1867, Canada was a union of English and French North Americans, but over time, the ethnic composition of the Canadian population diversified into what is today one of the world's most multicultural societies. According to the CIA Factbook (2023), the ethnic makeup of Canada in 2021 was Canadian 15.6%, English 14.7%, Scottish 12.1%, French 11%, Irish 12.1%, German 8.1%, Chinese 4.7%, Italian 4.3%, First Nations 1.7%, Indian 3.7%, Ukrainian 3.5%, and Metis 1.5%. It should be noted that those identifying as either First Nations or Metis are now just over 3% in a country of which they were the original inhabitants and in whose territories all Canadians now live.

English and French are the official languages of Canada. Just over 87% of Canadians identify English as their language and just over 29% identify French as theirs (CIA 2023). As shown in the World Bank County Profile (Table 5.1) that follows, Canada is a developed country that boasts one of the highest standards of living in the world. It ranks 26/166 countries of the world on

J. Singh, *Sustainability and the Sustainable Development Goals in Cultural Clusters*, Sustainable Development Goals Series, https://doi.org/10.1007/978-3-031-70618-9_5

Table 5.1 Country profile: Canada

	1990	2000	2010	2021
World view				
Population, total (millions)	27.69	30.69	34	38.23
Population growth (annual %)	1.5	0.9	1.1	0.6
Surface area (km^2) (thousands)	9879.80	9879.80	9879.80	9879.80
Population density (people per sq. km of land area)	3.1	3.4	3.8	4.2
Poverty headcount ratio at national poverty lines (% of population)	–	–	–	–
Poverty headcount ratio at $2.15 a day (2017 PPP) (% of population)	0.2	0.2	0.2	0.2
GNI, Atlas method (current US$) (billions)	571.57	698.13	1512.74	1862.22
GNI per capita, Atlas method (current US$)	20,640	22,750	44,490	48,720
GNI, PPP (current international $) (billions)	540.03	877.36	1336.63	2015.13
GNI per capita, PPP (current international $)	19,500	28,590	39,310	52,720
People				
Income share held by lowest 20%	7.7	7.1	7.2	7.4
Life expectancy at birth, total (years)	77	79	81	83
Fertility rate, total (births per woman)	1.8	1.5	1.7	1.4
Adolescent fertility rate (births per 1000 women ages 15–19)	28	17	14	7
Contraceptive prevalence, any method (% of married women ages 15–49)	–	80	–	–
Births attended by skilled health staff (% of total)	99	99	99	98
Mortality rate, under-5 (per 1000 live births)	8	6	6	5
Prevalence of underweight, weight for age (% of children under 5)	–	–	–	–
Immunization, measles (% of children ages 12–23 months)	89	96	90	90
Primary completion rate, total (% of relevant age group)	–	98	–	–
School enrollment, primary (% gross)	104.3	100.8	98.6	101.8
School enrollment, secondary (% gross)	100	102	102	115
School enrollment, primary and secondary (gross), gender parity index (GPI)	1	1	1	1
Prevalence of HIV, total (% of population ages 15–49)	–	–	–	–
Environment				
Forest area (km^2) (thousands)	3482.70	3478.00	3473.20	3469.30
Terrestrial and marine protected areas (% of total territorial area)	–	–	–	10.8
Annual freshwater withdrawals, total (% of internal resources)	1.6	1.5	1.3	1.3
Urban population growth (annual %)	1.5	1.4	1.3	0.7
Energy use (kg of oil equivalent per capita)	7630	8265	7789	–
CO_2 emissions (metric tons per capita)	15.15	16.76	15.79	13.6

Electric power consumption (kWh per capita)	16,167	17,037	15,270	–
Economy				
GDP (current US$) (billions)	596.09	744.77	1617.34	2001.49
GDP growth (annual %)	0.2	5.2	3.1	5
Inflation, GDP deflator (annual %)	3.4	4.3	2.8	8.2
Agriculture, forestry, and fishing, value added (% of GDP)	–	2	1	2
Industry (including construction), value added (% of GDP)	–	30	26	24
Exports of goods and services (% of GDP)	25	44	29	31
Imports of goods and services (% of GDP)	25	39	31	31
Gross capital formation (% of GDP)	21	21	23	24
Revenue, excluding grants (% of GDP)	19.7	20.9	16.9	19.1
Net lending (+)/net borrowing (–) (% of GDP)	–4.5	2.2	–2.2	–3.2
States and markets				
Time required to start a business (days)	–	4	6	2
Domestic credit provided by financial sector (% of GDP)	100.2	110.7	–	–
Tax revenue (% of GDP)	14.1	14.9	11.7	13.3
Military expenditure (% of GDP)	2	1.1	1.2	1.3
Mobile cellular subscriptions (per 100 people)	2.1	28.4	76	88.1
Individuals using the internet (% of population)	0.4	51.3	80.3	92.8
High-technology exports (% of manufactured exports)	–	–	15	15
Statistical capacity score (overall average) (scale 0–100)	–	–	–	–
Global links				
Merchandise trade (% of GDP)	42	70	49	51
Net barter terms of trade index (2000 = 100)	–	100	115	100
External debt stocks, total (DOD, current US$) (millions)	–	–	–	–
Total debt service (% of exports of goods, services and primary income)	–	–	–	–
Net migration (thousands)	145	202	228	195
Personal remittances, received (current US$) (millions)	–	555	1199	832
Foreign direct investment, net inflows (BoP, current US$) (millions)	8402	68,303	29,715	64,727
Net official development assistance received (current US$) (millions)	–	–	–	–

World Bank (2023)
Source: World Development Indicators database
Country: Canada
Data from database: WDI Central

progress towards the achievement of the SDGs. Its score of 78.5% shows its progress towards the 2030 deadline (Sachs et al. 2023).

5.1 Narrative Themes from Canada Leaders

The themes that emerged from interviews with the Canadian leaders are presented in this section and are summarized in Table 5.2. These themes are described in the following subsections and are summarized at the end of the section.

5.1.1 Closing the Wealth Gap

Both Canadian political and business leaders believed that closing the gap between Canada's rich and poor might help economic sustainability. Participant CP1 compared the wealth gap in Canada and the USA, and claimed that conflicts existed in the USA due to the large gap between the wealthy 1% and the poor. Participant CP1 claimed:

> *To the extent that wealth gets held by just a few people or a small 1% of the population; and that 1% can have a significantly disproportionate effect on political outcomes that naturally affect likely the entire population. You'll see more discontent. You'll see more revolt to decisions and people demanding other things. There are definitely problems that would happen. I don't think it's sustainable. To get to the heart of it, I don't think it's sustainable the way it is.*

Table 5.2 Major themes from Canada

Pillar of sustainability	Business leaders	Political leaders
Economic	Closing the wealth gap	Closing the wealth gap
	Adapting to change	Making and adapting to changes
Social	Employment equity and inclusion	Employment and local culture
Environmental	Efforts of the private sector	Government taking action
	Government policies	

Asked whether the situation was any different in Canada, Participant CP1 responded:

> *Uh, I think that there's probably some mirrors in Canada and between Canada and the United States. I think our system may build in different buffers that don't allow it to play itself out so raw. You know what I mean?*

Participant CP2 mentioned that Canada was not experiencing income inequality; however, in other countries, the private sector, particularly "big corporations" was highly focused on profit among other things. Participant CP2 stated:

> *They're not overpaid. But, they're not underpaid. And I think it's a fair wage, and I think if people use that four letter F word, Fair, I think if we treat people fair, that would take a lot of this inequality out. But, sadly, as the big corporations, mainly, they're there for the bottom dollar. The profit. But, if we went back to the old ways, where Mr. Ford said if they were paid a fair wage, they could afford to buy the vehicles we manufacture. That's being done away with, sadly.*

Political leaders generally believed that the private sector affected the economy. Participant CP3 explained:

> *And it so it's an investment to reduce competition for the future, so I mean it's a loss leader for some businesses because it's more expensive for them to compete versus buying their competitor and then eliminating their competitor. It's a mathematical equation. The unfortunate thing is that if there are people in it, it implodes the capitalistic model of the free enterprise, and the better idea comes and all that kind of stuff. It does, which you know. So, I think we're in a unique situation. Economic contributions through private equity for instance; we don't even know who owns them, what their motivation is. So I think we're in a very unique situation, which is hard to determine where it goes, but I think it does create the underpinnings of greater social strife and greater economic uncertainty because there is no general model based on this.*

However, business leaders generally perceived that government regulation, through taxation, was the solution to the wage gap. Participant CB5 shared:

> *So there's kind of a piece of me that says, "Yeah you need taxation." That helps to make sure that we take of the people who through either social circumstances made it more difficult for them to get ahead, or just simply to do experiments with living wage and figure out what really works out.*

Participant CB2 predicted trouble ahead if inequality is not addressed. This participant is widely regarded as one of Canada's top CEOs:

> *Yeah, I do. I do. I hate the new tax plans of the United States. I don't believe in trickle-down economics. It's never been proven. Typically, people will just buy back stock or retire debt, and the stock will go up, and the CEOs get more money. Presidents are being paid way too much money. The average president in the United States or CEOs are like $10 million. Tax rates have never been lower, so what's happening to that? They're cutting back stuff. Parks are not being taken care of. Schools, schools in the US are in horrible shape. Libraries are being closed. All of that stuff, it negatively impacts the sustainability of the communities. Eventually these people, I think, are going to rise up. And they've got 300 million guns there. There could be another Civil War down there.*

5.1.2 Making and Adapting to Change

Both political and business leaders perceived that change contributed to economic sustainability. While business leaders supported adapting to changes, political leaders emphasized taking action to make changes as well. Some participants mentioned the impact of technology on the economy. Participant CB3 mentioned:

> *It will have big influence. It already is, but it'll have big influences in the future in how we perform our work, and the need for labour, the types of labour, the shift in the roles in the economy, whether it's an intelligence or robotics, et cetera. But just purely from saving from an environmental perspective, the digitization and reduction of paper use and lights and energy is really, really important.*

Similarly, Participant CP3 claimed that Canada was experiencing a "digital economy" in which hard assets were no longer the sole basis of economic growth. The participant shared:

> *Data are now gonna be stored in the cloud or in a server somewhere else where ownership of that material's in a transition.*

In adapting to change, CB4 reiterated the significance of "innovation" and explained:

> *We cannot do what we were doing ten years ago and won't do ten years from now what we're doing today. Sustainability is going to change, and people who don't change, they won't stay.*

Participant CB5 saw adaptation to change as perhaps including shifting global economic domination:

> *I look at the economic pillar, and I say North America, Western Europe are not necessarily going to be in the position economically they have been for maybe the last 50 years. So it will change, and I'm not sure that it's right to expect that it won't. Can I guarantee that that's gonna be the same thing for my kid? Maybe. Overall, for all of children, I think it's gonna be challenging. Just like other people had to perhaps consider immigration to another country, then my kids or my grandchildren might have to think about learning Mandarin and going somewhere else. The world has changed, I don't know if it's really feasible to say that it's perfectly sustainable, you'd like to hope. I think there's certain things that will continue, but it's like how long is reasonable? I can probably predict a lifetime, and say, "Yeah, more or less." But two lifetimes, then it gets a little blurrier. Even one, I think about in the time that I've graduated from university what's changed; it's pretty astronomical.*

5.1.3 Employment Equity and Inclusion

Both political and business leaders perceived that employment equity and inclusion of all genders and races influenced social sustainability. Both groups of participants claimed that women in Canada experienced employment and income inequity due to gender roles. Participant CP1 explained that women's roles in marriage, childbirth, and child-rearing affected their chances for employment. The participant stated:

> *I think that as long as women continue to give birth, they will be at a disadvantage in terms of their male counterparts, even if they're equally as smart; because taking yourself—you want to spend time, and it's right to spend time raising your children and taking a year off—but that is difficult for*

employers. Unless you're a large government here, it's difficult for a lot of employers to deal with, especially if someone's gonna be off for two or three years. Now in the context of a 30-year career, it probably isn't that big of a deal; but I've seen it happen. I've seen women be picked over because they're like, "Well, she's 28 or 29. She's probably gonna have a few kids. We need someone who's gonna ..."

And so I would say that it will be—is it sustainable? Yeah, it probably is sustainable. Is it fair? No. It's not fair.

The participant, CP1, an elected official, is of the view that although unfair, gender discrimination can continue indefinitely. The participant was essentially saying that there are significant negative consequences in the workplace for women who choose to have children and that the system could and does tolerate such unfairness.

The business leaders acknowledged the issue of employment and income inequity experienced by women; however, the participants claimed that such issues should not exist. The business leaders shared that hiring and providing wages to women should be performance-based. Furthermore, hiring women should be considered as an advantage, as women introduced diversity of opinion in the workforce. Participant CB1 stated:

So that if we have five of us who are all sitting around the table, and we all think the same, it's not good. It's a group think. So a lot of the research would show that when you have a woman on the board, you get more diversity of opinion and maybe you get some softer positions. But then at the same time, it's not just women. Maybe it's a Black person, it's an African-American, it's a Muslim, or whatever. So diversity of view. But just to say that your board should be made up of the same basis as the split in the population of the country, no. Because as a shareholder, I'm investing in your particular company, and I have the right to vote for who I want.

Some participants also mentioned eliminating racial bias for social and economic sustainability. Participant CB5 noted that racial bias is inimical to sustainability:

I don't think so because part of the issue of keeping your GDP growing is to have full participation in the workforce. So if you're not accessing a portion of your workforce by making it more difficult for them to work, then you're not able to continue to grow your GDP to the level you'd like to, which then perpetuates the issue of who's gonna be the leader in the world in the future. So if you get an aging population, and you're not making use of everyone you got there, then that's a problem. That's really what it comes down to for me. Are there social issues? Absolutely, but I just don't think it makes sense. It's basically saying I've chosen not to utilize a tool that I could use to make the economy better. Again you've got, like any other situation, the people in power don't see a need to change, and so unfortunately, I think you have to sometimes put rules in place. I'm in favour of things like board representation in particular, or in the US, I'd say some sort of universal maternity leave that makes it at least reasonable for women to be able to come back to work or to participate at a higher level. It's still tough. I always said I was lucky in that [when] I came out of business school. I probably was about the right point in time in terms of overt sexism was diminishing, but it still exists, and it's not overt anymore, it's about a group of people who are comfortable with the same group of people, so they continue to hire the same group of people. That's beyond women or anyone else either.

5.1.4 Efforts of Private Sector

Business leaders perceived that environmental sustainability was a joint effort of government and the private sector. The perceived contribution of the government will be presented in the next section. The majority of the business leaders believed that taking small actions such as building a "green" office, reducing production waste, and encouraging employees to save electricity may contribute to environmental sustainability. Participant CB1 claimed that such acts were not only helpful to the environment, but also saved on costs. The participant narrated:

Well, you know when you look at buildings and the LEED gold standard. So in our case, we moved down to the waterfront, and we had the opportunity to get a LEED gold building. So you're in a situation where the lights go off when you leave the room, so you reduce your energy consumption. When we moved all of our employees into the building, what we did was we gave everybody their own kind of personalized water bottle and said, "Maybe you should stop drinking the bottled water." And some little things like that. And then you'll have people put in rules—your car can't idle out front and small things like that. Or certainly waste, with separation and the recycling. Do you really need to provide all of that extra packaging? And what happens is that the corporation says, "I'm doing the right thing from an environment point of view, but oh, you know what? It's also saving me costs."

Participant CB3 added that encouraging employees to be environmentally conscious may produce a "cumulative impact" towards the environment. The participant shared:

So, these little steps of reducing our paper, reducing our footprint, thinking about how we work and do work and how we communicate with each other and where we work, really important. And as well, we talked earlier about transparency around is it sustainable to continue to work in the way that we do? Is inequality sustainable? I think what we're seeing examples now in policy just recently here in Ontario with the Pay Transparency Act.

Participant CB2 was forthright in condemning the skeptics on the need for more action to save the environment. He narrated:

Listen, I've spent three days with Al Gore. I went down not needing to be convinced on climate change because I believed it's happening. I wanted to be educated on it, and whether you're at sea temperature or glaciers, it's proven empirically. Ninety-seven percent of science believe it. They're peer-reviewed, studied, and the 3% who don't believe it, it's all anecdotal, and it's opinions.

Unfortunately, in the United States, whether it's evolution or climate change they say, "Well, we should teach both. We should teach the science and the opinion." No. Science is science. It's not opinion. You try to explain what science means. We have a major issue that's been getting worse and worse. The Paris Accord, which was signed two years ago, we've got to keep it to 1.5% climb or 2%. I'm very worried about doing that even though a 100-some odd cities signed it, and there seems to be some will for it. But then you've got somebody like Trump, who's doing anything he can do to dismantle it. Now then on top of that, you've got the C40 cities who are the cities saying, "Heck with you. We'll do it."

But I'm really worried. Even Canada, to be a little crude, I'll say they suck and blow. Trudeau says the right things, but we want to take all the dirty, dirty oil, one of the dirtiest oil, and we ship it to the West Coast. Did you see that tanker? Did you hear about that tanker crash in the China Sea? What would that do the West Coast of British Columbia? What's happening? We've been kind of ignoring this for a long time. Al Gore predicted it 12 years ago in An Inconvenient Truth, *and it's coming true. The scientists are all predicting it.*

5.1.5 Government Action

Both groups of participants perceived that government action contributed to environmental sustainability. The business leaders perceived that the government had failed to take adequate action to help the environment. Participant CB1 claimed that government officials were more focused on getting re-elected than on doing what was necessary. The participant explained:

Interestingly, I think that business has to do it. The politicians get too involved in it; in a sense that they're trying to protect certain areas which they probably shouldn't. But at the same time, I do believe that we should have some regulations because there are a lot of people who won't do the right thing ... If I have a little small company on the river, and I'm polluting into the river, and you're the really big company on the river and they say, "Oh, we're gonna put controls on you." Well if you're polluting the river, you're polluting the river. Now you might be small, but you're still polluting it. So I believe that there has to be some political regulation that's put in place, but at the same time, I really think that the politicians don't have the guts or the nerve to stick with it. 'Cause they're all focused on getting re-elected. And the corporations have to start to do the right thing, and a lot of them are.

The participant here is alluding to the view that perhaps business may be ahead of the government on environmental sustainability.

On the flip side, political leaders perceived that the government was attempting to reach environmental sustainability for future generations. Participant CP1 claimed that environmental sustainability involved long-term planning, especially in terms of the development of infrastructure. Furthermore, the participant also believed that the national government had the power to prevent environmental destruction and climate change. However, as narrated by Participant CP3, action on the environment lacks urgency:

So our environment, we're on borrowed time. Not only as a country, but I think collectively as a species. I was talking to Daniel [a Member of Parliament] the other day, and I was saying that he should revive his father's question [who was also a Member of Parliament], who asked the first question to the House of Commons on climate change.

The participant thought that the same question on climate change a generation later was still appropriate as little had changed for the better.

5.2 Summary of Themes from Interviews with Canadian Leaders

The themes that emerged from Canadian political and business leaders were:

1. Closing the wealth gap
2. Making and adapting to changes
3. Employment equity and inclusion
4. Effort of private sector
5. Government action

The Canadian interviewees thought that economic sustainability may be reached through closing the wage gap between the rich and the poor. In addition, the government initiating changes, and implementing them, while the private sector adapted and complied with the changes was also perceived to increase economic sustainability. The leaders generally believed that technology contributed to the changes in the economy. They also thought that social sustainability was associated with employment equity and inclusion. The majority of political and business leaders recognized the issues of the inequity experienced by women and minority groups. Nonetheless, business leaders shared that the inclusion of all genders and races in the workforce did not only contribute to social sustainability but also to economic growth. Lastly, environmental sustainability was perceived to be influenced by the private sector and the government. The federal government was seen to have the power to enact legislation for the protection of the environment and the private sector was obligated to comply with government policies and may make additional efforts such as encouraging employees to be environmentally conscious. The business leaders made the business case for sustainability and shared that environmental consciousness appeared to be not only good for the environment but also for reducing the expenses of the company.

References

CIA (2023) The World Factbook: Canada. From https://www.cia.gov/library/publications/the-world-factbook/geos/ca.html. Accessed 19 July 2023

Sachs JD, Lafortune G, Fuller G, Drumm E (2023) Implementing the SDG stimulus. Sustainable Development report 2023. SDSN/Dublin University Press, Paris/Dublin. https://doi.org/10.25546/102924

World Bank (2023) Country profile Canada. From https://databank.worldbank.org/views/reports/reportwidget.aspx?Report_Name=CountryProfile&Id=b450fd57&tbar=y&dd=y&inf=n&zm=n&country=CAN. Accessed 31 July 2023

6 Case Study 3: India

Abstract

This is the third of five chapters that analyze the interviews conducted with business and political leaders in selected countries. In this chapter interviews with leaders in India are examined. The chapter begins with background information on the country's history, economy, people, and environment. This is followed by the identification and discussion of themes on sustainability that permeate the interviews. The themes that surfaced in the Indian interviews were efforts of the private sector, sustainability is a responsibility, employment equity, changes brought by technology, education of the people, everyone's responsibility, and government policies.

India boasts one of the world's earliest civilizations, but as described in the title of one of V.S. Naipaul's books, it is now a wounded civilization (Naipaul 1979). It is wounded in the sense that although a vast nation with a rich culture and a developing economy, it suffers from severe inequalities—economic and otherwise. Moreover, a fallout of a fast-growing economy has been environmental degradation, such that some of India's large cities are now the most polluted in the world. However, as the world's largest democracy and with the largest population of more than 1.4 billion, India is a very important player in world affairs. All of this makes India a critical nation in the achievement of the United Nations' Sustainable Development Goals.

India, situated in Southern Asia, is more than three million square kilometers, making it the eighth-largest country in the world. Its land neighbors are Burma and Pakistan; the latter having been carved out of India at Independence in 1947. The CIA Factbook lists deforestation, soil erosion, overgrazing, desertification, air pollution from industrial effluents and vehicle emissions, and water pollution from raw sewage and runoff of agricultural pesticides among a long list of environmental issues facing the country (CIA 2020, 2023).

The diversity of the Indian population is reflected in ethnicity, religion, and languages spoken: 72% of the population is classified as Indo-Aryan; 25% as Dravidian; 80% as Hindu; and 14% as Muslim. Dozens of languages are spoken with Hindi and English being among the most prevalent (CIA 2020). The median age of this rapidly changing nation is just below 29 years.

Despite its diversity and associated challenges, India has remained a parliamentary republic with a federal government and 28 state and 8 union territory governments. The current Prime Minister is Narendra Modi of the Bharatiya Janata Party (BJP).

J. Singh, *Sustainability and the Sustainable Development Goals in Cultural Clusters*, Sustainable Development Goals Series, https://doi.org/10.1007/978-3-031-70618-9_6

PM Modi, who has been in office since 2014, is regarded as an economic reformer and has been valiantly struggling to move India out of poverty and with some success. India's GDP of approximately $9.5 trillion (Purchasing Power Parity) makes it the world's third-largest economy, but its per capita GDP of $7200 (PPP) is rather modest (CIA 2020). More details on the economy, people, and environment of India are provided in Table 6.1. India's SDG progress score is 63.45/100, which ranks as 112/166 among countries assessed. The country is on track to achieve only about 34% of the SDG targets and is making limited progress on a further 43% but worsening on 33% (Sachs et al. 2023).

6.1 Narrative Themes from India Leaders

The themes that emerged from the interviews of Indian leaders are presented in this section and are summarized in Table 6.2. These themes are described in the following subsections and are summarized at the end of the section.

6.1.1 Efforts of the Private Sector

As business leaders in India, some of the participants believed that profitability was the priority of any business. Participant IB3 argued that this was not sustainable but must be pursued in conjunction with social and environmental goals. The participant explained:

> *So now the economic aspect is the one which starts touching us because the business has to be economic and sustainable. So for me, it's like a mix of all these things. So social and environmental aspects we can take care of it, provided we are having economically sustainable business.*

The participant further stated that his business was making an effort to create business plans that would positively impact the economy, society, and environment. The participant showed a video about the business, which highlighted how the company provided a sustainable service. In the words of the participant:

> *I think it's difficult to take the view which an environmentalist will take, or which a politician will take about sustainability because we are in the business, and the business has to be viable. The only motive is not making profits, but it has to be viable, because if you're not viable, we won't survive for long. So we can do some business for the short-run, but it won't be viable for the long-run. So in that context, if you look at sustainability for us—the sustainability from this economic, social, or from the environmental perspective—all three perspectives become something where people at large are willing to buy into that business model. Because essentially, if we don't have a product or a service with which a large number of people feel comfortable, I don't think we are doing too much of service towards sustainability. We can do experimentation, we can do pilot projects, but it won't take us very far.*

Participant IB1 believed that economic sustainability was gradually improving. The continued existence of India despite extreme poverty was perceived to be evidence that the economy was doing well. He attributed this to the efforts by the private sector in helping to keep the economy afloat. Participant IB1 stated:

> *But the way we had the growth from the last 15 years, it proves that sustainability is there. On all fronts. On all fronts. Otherwise we would not have been what we are, despite all sort of negativism that has come in each and every field. We say that education has helped. We are one of the poorest countries in the world. Economically we are backward; environmentally, there's a lot of pollution and everything. But despite all of that, we are what we are today. And some say that we are the fastest-growing economy in the world. Fastest-growing economy itself proves that we are sustainable. Because tomorrow is better than today.*

For this Indian business participant, sustainability is defined by economic growth. As he explained, while the environment may be deteriorating and social problems fester, if the economy

Table 6.1 Country profile: India

Country profile				
	1990	2000	2010	2021
World view				
Population, total (millions)	870.45	1059.63	1240.61	1407.56
Population growth (annual %)	2.1	1.8	1.4	0.8
Surface area (km^2) (thousands)	3287.30	3287.30	3287.30	3287.30
Population density (people per sq. km of land area)	292.8	356.4	417.3	469.7
Poverty headcount ratio at national poverty lines (% of population)	45.3	–	21.9	–
Poverty headcount ratio at $2.15 a day (2017 PPP) (% of population)	47.6	–	22.5	10
GNI, Atlas method (current US$) (billions)	334.59	467.08	1505.74	3019.86
GNI per capita, Atlas method (current US$)	380	440	1210	2150
GNI, PPP (current international $) (billions)	1034.51	2188.20	5173.87	10,163.36
GNI per capita, PPP (current international $)	1190	2070	4170	7220
People				
Income share held by lowest 20%	8.9	–	8.1	8.1
Life expectancy at birth, total (years)	59	63	67	67
Fertility rate, total (births per woman)	4	3.4	2.6	2
Adolescent fertility rate (births per 1000 women ages 15–19)	122	101	39	17
Contraceptive prevalence, any method (% of married women ages 15–49)	45	47	55	67
Births attended by skilled health staff (% of total)	34	43	52	89
Mortality rate, under-5 (per 1000 live births)	127	92	58	31
Prevalence of underweight, weight for age (% of children under 5)	52.8	45.3	–	31.5
Immunization, measles (% of children ages 12–23 months)	56	56	82	89
Primary completion rate, total (% of relevant age group)	64	71	93	97
School enrollment, primary (% gross)	91.4	94.3	109.1	102.1
School enrollment, secondary (% gross)	37	45	63	78
School enrollment, primary and secondary (gross), gender parity index (GPI)	1	1	1	1
Prevalence of HIV, total (% of population ages 15–49)	0.1	0.6	0.3	0.2
Environment				
Forest area (km^2) (thousands)	639.4	675.9	695	721.6

(continued)

Table 6.1 (continued)

Country profile				
	1990	2000	2010	2021
Terrestrial and marine protected areas (% of total territorial area)	–	–	–	4.4
Annual freshwater withdrawals, total (% of internal resources)	34.6	42.2	44.8	44.8
Urban population growth (annual %)	3.1	2.6	2.5	2.1
Energy use (kg of oil equivalent per capita)	351	416	559	–
CO_2 emissions (metric tons per capita)	0.65	0.89	1.34	1.58
Electric power consumption (kWh per capita)	273	393	637	–
Economy				
GDP (current US$) (billions)	320.98	468.39	1675.62	3150.31
GDP growth (annual %)	5.5	3.8	8.5	9.1
Inflation, GDP deflator (annual %)	10.7	3.6	10.5	8.5
Agriculture, forestry, and fishing, value added (% of GDP)	28	22	17	17
Industry (including construction), value added (% of GDP)	27	27	31	26
Exports of goods and services (% of GDP)	7	13	22	22
Imports of goods and services (% of GDP)	8	14	27	24
Gross capital formation (% of GDP)	27	26	40	31
Revenue, excluding grants (% of GDP)	12.4	11.7	13.2	13.2
Net lending (+)/net borrowing (–) (% of GDP)	–3.4	–3.8	–3.5	–2.3
States and markets				
Time required to start a business (days)	–	–	33	18
Domestic credit provided by financial sector (% of GDP)	–	–	–	–
Tax revenue (% of GDP)	10	8.8	10.4	12
Military expenditure (% of GDP)	3.1	2.9	2.9	2.7
Mobile cellular subscriptions (per 100 people)	0	0.3	60.6	82
Individuals using the internet (% of population)	0	0.5	7.5	46.3
High-technology exports (% of manufactured exports)	–	–	8	10
Statistical capacity score (overall average) (scale 0–100)	–	–	81	77
Global links				
Merchandise trade (% of GDP)	13	20	34	31
Net barter terms of trade index (2000 = 100)	86	100	93	105
External debt stocks, total (DOD, current US$) (millions)	83,472	101,131	290,428	612,866
Total debt service (% of exports of goods, services and primary income)	33	17.1	6.8	7.3
Net migration (thousands)	108	–150	–380	–302

Country profile				
	1990	2000	2010	2021
Personal remittances, received (current US$) (millions)	2384	12,883	53,480	89,375
Foreign direct investment, net inflows (BoP, current US$) (millions)	237	3584	27,397	44,727
Net official development assistance received (current US$) (millions)	1395.00	1383.40	2831.30	3118.70

World Bank (2023)
Source: World Development Indicators database
Country: India
Data from database: WDI Central
Last Updated: 07/25/2023

Table 6.2 Major themes from India

Pillar of sustainability	Business leaders	Political leaders
Economic	Efforts of the private sector	Sustainability is a responsibility
Social	Caste discrimination	Employment equity
	Employment of women	Culture and religion
	Wage and opportunity gap	
Environmental	Changes brought by technology	Everyone's responsibility
	Education of the people	Government policies
	Experiences in a developing country	

continues to grow, then that means that current policies and practices are sustainable, and therefore, there is sustainability.

6.1.2 Sustainability Is a Responsibility

Political leaders in India generally perceived that sustainability, particularly economic sustainability, was a responsibility. Economic sustainability was perceived to require planning and conscious effort on the part of political leaders. Participant IP5 shared:

> *Yes, India needs help implementing the infrastructure sector. If you go to see, there are a lot of accidents which take place on the roads. These are the basic things. We need good road infrastructure which will again reduce your pollution levels. See, if you have a good road network, it will. It gives impetus to the economic development. And we need sustainable economic development. A place like this state should attract industries more of a service kind or more of [an] assembly kind. Not the hard-core chemical manufacturing or hard-core raw material manufacturing. So we are blessed with ports, we are blessed with airports, we are blessed with a fairly good energy power sector. We have good renewability of water. So it is up to us, politicians, the thinkers in this state, or the guiding people of this state. We should be willing to put good road networks, basic infrastructure in place.*

Participant IP6 perceived that construction was being done all over India as the country develops. As such, the participant believed the construction industry needed overarching values to guide decision-making in order to achieve economic sustainability. Participant IP6 stated:

> *So sustainability for us is an idea which should be at the back of your mind when taking any decision—whether it concerns our studies, whether it concerns our purchases, whether it concerns construction, whether it concerns teaching in the classroom, whether it concerns dealing in things with the masses, and whether it has any concern with the decision-making that we do for public policy purposes.*

Participant IP6 continued to describe a stewardship relationship with the environment:

> *When we talk of sustainability in India, the first thing that comes to us is sustainability means living with nature. Sustainability to us means living not for today, but living for future. Sustainability for us means not only profits, not only output, not only productivity, not only willing to gain, but sustainability with us means utilization of resources in such a way that we don't dilute the resources that were created for millions of years.*

Participant IP2 believed that India lacked significant government regulations for the development of a sustainable economy. The participant also perceived a lack of consequences for people unconcerned about sustainability. Participant IP2 cited the example of a large sum lost to the state because of illegal mining:

> *Hopefully it will be recovered. Then you'll talk of environmental damage. In such a short span of time, which could've occurred in another 50 years, but at the same time, replenishment would have taken place. The question is less to do with enforcement, more to do with development of the culture to obey the laws. Punishment is required as a deterrent. You have to show some examples, that violation will not be tolerated, then automatically things fall in place. But, who is enforcing? How many vehicles are checked? See, the difference between them and us is this. They're: "I will not wait for somebody to catch me and then I will do it. I will do it in time … because the penalties are huge!"*

6.1.3 Employment Equity

Political and business leaders in India perceived that employment equity influenced social sustainability. Political leaders believed that culture and religion generally influenced employment equity. Participant IP6 articulated:

> *So this social equity problem, in terms of employment, you'll find that both from demand side and from supply side. We are taking, I think, a number of actions and it is going alright. I think it is more now political. I think it is more a political dimension. I would say, but you find a unison I think among people, and that kind of discrimination which used to be there, that "you are you" and "we are we", that is going away, because this educational system infrastructure, you find accessibility and all it's there for the majority of the people.*

Participant IP6 argued that the educational system was helping to break down barriers to the employment of women. He suggested:

> *If you take this institute where they have six hundred students, you'll find that three hundred of them are girl students. And we find that our examination, public examinations, they're taught by girls; so far, they are taught by, so far, boys and girls from different sections of society. Our management schools, technical schools, technology schools, you find them, both boys and girls who are not from upper sections. They are making their way there, and they start performing.*

Business leaders similarly perceived that caste discrimination and the patriarchal society influenced employment equity. Indian leaders also believed that India did not only experience a wage gap but also an opportunity gap. An illustration of this is the caste system, whereby people from lower castes tend to receive less opportunity, and therefore, less wages. Participant IB3 shared:

> *Inequality is increasing. It's definitely increasing. I remember seeing some articles about that. But the challenge is that: Is your base population also able to have a decent lifestyle? So ... again, boils down to their basic needs. Are you able to provide people at least some of the reasonable standard of living. Are they out of the slums? Are they having access to clean water? Energy? Are they having any access to basic shelter and food? That's where it starts to grow there, so it becomes the same. Inequality will be there. In any capitalist society, it will be there. We are not a communist society, so we cannot expect everyone to have the same bread every morning. People will have different things.*

Although arguing that there is generally no religious discrimination in employment, Participant IB4 suggested that perhaps under certain circumstances there is religious discrimination. He elaborated as follows:

> *See, I'm from Hyderabad, which is a place full of resources. I'm a Hindu, and a lot of India, all kinds of businessmen in all religions. I don't think religion has ever played a part in any of our business meetings as far as giving equal opportunities goes. I would like to get the best guy. If he's going to be a Muslim or a Christian or a Jew, I don't care. If he's going to be an affordable price and he's the best guy, I will do it. And that's what most of the private sector firms will do. Most of the private sector people would like to get the best resource for the price that you are willing to pay. Religion is not going to play a part. I do not think it's going to play a part. I know this for a fact with a lot of Hindu companies, and lot of Muslim companies, too. But what happens is, in all these places, there will be that element of the small ... maybe I might, when I am demanding a particular level of competence in some other, everybody, maybe I might give a 5%, 10% discount for somebody in my place if he is not going to be detrimental to my good. That driver is always there.*

6.1.4 Changes Brought by Technology

Some business leaders believed that technology might improve environmental sustainability in India. Participant IB3 shared that having a "technology-driven" company contributed to the distribution of sustainable products. The participant shared:

> *My company has an in-house expertise of providing environmental solutions. Our more than 75 people come from world-class utilities, with vast experience in designing, buildings, and operating solar plants. We specialize in designing the rooftop power plants, and it's structured to ensure most optimum generation by ensuring the highest quality of health and safety practices.*

Similarly, participant IB4 claimed that having a business in the field of technology led to encounters with highly educated people, suppliers, and customers alike, who valued sustainable materials. The participant explained the role of adaptation to changing circumstances:

> *To sustain in that, I also had to be knowledgeable. That is also other essential teamwork we required. Apart from being politically correct and quality being good and all that, we also had ... our data, as the customer basis goes, we were out of education. If I were to continuously show myself as a better person I had to talk to them about technology and technical things which are maybe up-to-date.*

6.1.5 Education of the People

Some business leaders emphasized the significance of education to environmental sustainability. Participant IB1 claimed that education influenced the population in terms of choosing to stay in cities and continue to develop the cities. Participant IB3 shared:

> *Yeah. I think the important part is that as people become more educated. It's a little bit a part of the education. And as people become more educated and understand the importance of a cleaner environment, cleaner water, cleaner surroundings, then these things will improve. And that's where I think the government is doing a reasonable job of promoting those things. That, "Let's do it like this. Let's try to use whatever the food waste as a compost."*

6.1.6 Everyone's Responsibility

In terms of environmental sustainability, the majority of Indian political leaders believed that everyone had roles to play. Everyday actions could impact the environment. For instance, participant IP6 stated:

> *But I would say that in my view I think the best way, there are two ways to do that. One is supply.... So if you think of India, for thousands of years I was searching the first thing you should do is to treat this in a very big way in our daily activities, where our children—children are the most important persons—like the grown person should have the same thinking that don't waste power. That don't waste light, that don't create pollution, that don't create garbage unnecessarily. Keep your house clean. That road, it's not there because of the government. If litter is there, it's your responsibility also to help. Don't blame that unless government person comes and lifts it, you'll just be seeing the rubbish, but you won't extend your hand and take it from the other side.*

Some Indian political leaders believed that religion guides people in how they treat the environment. Hinduism was believed to teach people to care for the environment. Participant IP3 narrated:

> *Whenever a change happens in the nature, the human being has a big role in it. He changes the nature, therefore the change comes from the nature. We want to communicate that ... let's be a sustainable development but do not harm the nature. Development is a continuous process and keeps moving. According to Hindu mythology, nature destroyed everything, only "Manu and his wife" lived and the rest of the earth would have gone. Now again whole universe has built and with huge population. Continuous development would allow, but it should follow the principle of sustainability, otherwise nature will do balancing by its own.*

Participant IP3 continued by saying that while human beings cannot control nature, human beings can save nature. The participant emphasized:

> *So I say nature is changing and nature has a power to change in everything in environment. Man can't control every change made by the environment, but yes, in order to save the nature, we can plant trees to mitigate degradation.*

Environmental sustainability was also believed to be influenced by obeying the law. While the Indian government has passed various pieces of legislation on environmental sustainability, enforcement of the law was seen to be not very effective. However, participant IP2 believed that it was everyone's responsibility to obey the law and conserve nature. The participant stated:

> *See, it's not a question of enforcement. It is a culture that has to develop that I have to obey the law, which we have ... you know, some have gone wrong in this aspect, so that has to be emphasized again. It has to be done by doing it from the grassroots, from the schooling level, understand that we are to be obedient and embrace change.*

6.1.7 Government Policies

Concerning the previous theme, the Indian political leaders believed that principles of Hinduism guided policy-making on environmental sustainability. Participant IP4 mentioned:

> *In the Hindu religion, we believe that we should protect the environment. We teach our children this, and that's part of how we live. We believe that trees have life, everything has life, so you can't destroy them, because if you destroy the natural environment, you also will not survive When we are preparing food, for example, we prepare first for the cow and then the last is for us. These beliefs drive government policy. These beliefs drive government policy, yes, yes. These play a part in policy of the central government and of the state government. We think that's on the basis of these philosophies we can develop a sustainable society.*

Participant IP3 believed that India as a large country had an abundance of natural resources. Participant IP3 cited narratives from the "Holy Book Ramayana" of relevance to environmental policies. The participant shared:

> *For the Indian economy to be sustainable, the government has a lot of resources, because India is a country that has all the types of natural factors or geographical characteristics—like seas are there, there are mountains, there are plains, and deserts are also there. All kinds of natural characteristics in the world are available in India. India has a lot of resources; we do not have any problem to achieve economic sustainability. We have a lot of rich resources and available in abundance. There is great favour of nature to India; there are 70 minerals in this state alone, which will make India's economy remain stable. Apart from this, India extracts power from the waves of the sea, power from wind, thermal power, we have coal mines, solar energy, etc. All this makes Indian economy more stable and continues sustainability. Evidence taken from our* Holy Book Ramayana, *when Lord Ram went to Lanka, then built a bridge on the sea. Now USA wants to demolish that bridge as it obstructs the movement of ships in the sea. When India conducted a study, it was found that there is a Thorium in it, and Thorium more useful than Nuclear. This is the international conspiracy to destroy the Thorium and hamper the economic strength and stability of India. So I mean to say that if India will do it in Thorium, then [this] naturally will make our economy more sustainable. This is also one of the aspects which leads to sustainable Indian economy.*

6.2 Summary of Themes from Interviews with Indian Leaders

The themes that emerged from the Indian political and business leaders were:

1. Efforts of the private sector
2. Sustainability is a responsibility
3. Employment equity
4. Changes brought by technology
5. Education of the people
6. Everyone's responsibility
7. Government policies.

Concerning the economy, the interviewed Indian leaders perceived that the private sector played a significant role in sustainability. They contended that the private sector controlled the market. They also suggested that employment inequity involved caste discrimination, employment of women, wage and opportunity gap, and culture and religion. One participant mentioned the influence of a developing country on social sustainability, suggesting that developed countries had more opportunities. Environmental stewardship, technology, education, responsibility, and government policies were perceived to contribute to sustainability. In addition, the mention of alternative resources also underlined the influence of a developed country compared to the influence of a developing country on the ability to procure sustainable materials rather than emphasize industrial development. Some participants also cited the influence of Hinduism on the way locals treat the environment, and on the way political leaders develop environmental legislation.

References

CIA (2020) The World Factbook: India. From https://www.cia.gov/library/publications/the-world-factbook/geos/in.html. Accessed 25 July 2020

CIA (2023) The World Factbook: India. From https://www.cia.gov/library/publications/the-world-factbook/geos/in.html. Accessed 20 July 2023

Naipaul VS (1979) India: a wounded civilization. Penguin Books, New York

Sachs JD, Lafortune G, Fuller G, Drumm E (2023) Implementing the SDG stimulus. Sustainable Development report 2023. SDSN/Dublin University Press, Paris/Dublin. https://doi.org/10.25546/102924

World Bank (2023) Country profile India. From https://databank.worldbank.org/views/reports/reportwidget.aspx?Report_Name=CountryProfile&Id=b450fd57&tbar=y&dd=y&inf=n&zm=n&country=IND. Accessed 31 July 2023

7 Case Study 4: Sri Lanka

Abstract

This is the fourth of five chapters that analyze the interviews conducted with business and political leaders in selected countries. In this chapter interviews with leaders in Sri Lanka are examined. The chapter begins with background information on the country's history, economy, people, and environment. This is followed by the identification and discussion of themes on sustainability that permeate the interviews. The themes that emerged in the Sri Lankan interviews were long-term impact, the impact of globalization on economic sustainability, employment of women in a patriarchal society, lack of attention to discrimination against ethnic groups, the impact of globalization on social sustainability, government action on social sustainability, the impact of businesses on resources, locals lacking environmental consciousness, influence of Buddhist philosophy, government action, and experiences in a developing country.

Historically Sri Lanka has had various names. In ancient times it was known as just Lanka, meaning island, and in Hindu scripture it was here that Ravan took the Goddess Sita after kidnapping Lord Rama's wife. Lord Hanuman then performed miraculous feats in rescuing Sita and returning her to her husband. To this day, there are those who believe that the physical remnants of Hanuman's heroic feat can be seen in a possible structure jutting out of the strait between India and Sri Lanka. Sri Lanka means resplendent island and is the official name adopted by the country when it became a republic in 1972. For many years before that, it was known as Ceylon during the period of British colonial rule.

Whatever name it was known by Sri Lanka is indeed a resplendent island, physically beautiful and diverse, as are its 23,000,000 people, whose ethnic and religious diversity have been the cause of disunity and war. The most destructive of these conflicts was a civil war that started in 1983 and ended in 2009 with the defeat of a Tamil force seeking a separate state in the North of the country. Sri Lanka's former President, Gotabaya Rajapaksa, oversaw the effort that led to the defeat of the separatist Tamil uprising. As to be expected, the scars of the long conflict are yet to heal. According to the CIA World Factbook, the Sri Lankan population of 23.3 million includes ethnic diversity comprised of the following groups and in the indicated proportions: Sinhalese 74.9%; Sri Lankan Tamil 11.2%; Sri Lankan Moors 9.2%; Indian Tamil 4.2%; other 0.5% (CIA 2020; CIA 2023). The religious diversity of the country mirrors its ethnic diversity in that the Sinhalese are mainly Buddhist, the Tamils are

J. Singh, *Sustainability and the Sustainable Development Goals in Cultural Clusters*, Sustainable Development Goals Series, https://doi.org/10.1007/978-3-031-70618-9_7

mainly Hindu, and the Moors are mainly Muslim. Hence, the following religious diversity: Buddhist (official) 70.2%; Hindu 12.6%; Muslim 9.7%; Roman Catholic 6.1%; other Christian 1.3%; other 0.05% (CIA 2020). Conflicts have stifled the country's economic potential, and it is currently drowning in debt, mainly owed to China and used for infrastructural projects after the separatist war ended in 2009. Its foreign debt of US$55 billion is estimated to be close to 100% of Gross Domestic Product, an unhealthy debt burden that is unsustainable. In 2021, the country's per capita GDP was estimated to be $13,400 (purchasing Power Parity). Further statistics on the people, economy, and environment of Sri Lanka are shown in Table 7.1. The country has made modest progress towards the SDGs. Its score of 69.4 ranks 83/166 of the countries assessed in the latest report (Sachs et al. 2023). It is on track to achieve only 37.9% of the SDG targets while worsening on 33.3% of them (Sachs et al. 2023).

The perceptions of the leaders interviewed from Sri Lanka are presented in this section. The major themes that emerged from the Sri Lankan leaders' perception of sustainability are summarized in Table 7.2 and are described in the following subsections.

7.1 Narrative Themes from Sri Lanka Leaders

7.1.1 Long-Term Impact

Both Sri Lankan political and business leaders perceived that economic sustainability involved long-term profitability. However, business leaders perceived that in addition to profitability, long-term stability also drove economic sustainability. Participant SB3 explained:

> *If you're asking any corporate person, sustainability has a very strong correlation to financial section. Cash flow, and all that financial stuff. So the point I'm trying to drive here is: Economic sustainability may be financial stability, financial sustainability. So I don't see much of a difference in that context of the economic sustainability and the financial side.*

Participant SB1 believed that economic sustainability included stability of production and profit. The participant shared:

> *And what are they supplying in the way of goods for consumption? They are exhausting the natural resources of the country, of the world. To make these goods which they did and tell their people to consume. They make money; they're only just in it making some money. They should sustain at least and then can go. They want to make as much as possible, as many as possible, as quickly as possible, and they are off.*

7.1.2 Impact of Globalization on Economic Sustainability

Business leaders SB1 and SB2 believed that globalization had an impact on sustainability; however, SB3 perceived that sustainability began internally. Participants SB1 and SB2 had similar opinions on globalization, and shared that the more globalized Sri Lanka was, the more people engaged in consumerism. Participant SB1 illustrated his point with snippets from Sri Lankan history:

> *The foreigners who came in with their European concepts. Now, the Sinhalese king, most of the households did not have very much real furniture. They were not allowed to! He said no. And maybe the king had picked two people had more elaborate in nature of furniture, but most fellows hated the furniture. Hated the furniture. To that extent, we were not material. It came with those people and now, of course, is a change because all these people who had the village way of life are now exposed to these television programs, and it has created a big revolution in their aspirations. This is what happens in ......, and this is what I'm sure is happening in Colombo. We won't live a simple life anymore. And that is a very big disaster. Television has become, and yet, the people who foster more and more consumption and when they foster consumption, you got to go back, most of the advertisements are targeted to the women. We then, push the main focus. I want this; I want that.*

While Participant SB1 saw globalization as imposing unhealthy consumerism that is contrary to Sri Lankan traditions, Participant SB3 was of the view that there is community participation in fostering sustainability.

The participant explained that local communities played a part in sustainability:

Table 7.1 Country profile: Sri Lanka

Country profile				
	1990	2000	2010	2021
World view				
Population, total (millions)	17.2	18.78	20.67	22.16
Population growth (annual %)	1.2	0.6	0.9	1.1
Surface area (km^2) (thousands)	65.6	65.6	65.6	65.6
Population density (people per km^2 of land area)	274.3	299.4	329.6	354.3
Poverty headcount ratio at national poverty lines (% of population)	26.1	46.8	28.4	14.3
Poverty headcount ratio at $2.15 a day (2017 PPP) (% of population)	11.6	11	3.5	1
GNI, Atlas method (current US$) (billions)	8.05	16.41	49.28	88.58
GNI per capita, Atlas method (current US$)	470	870	2380	4000
GNI, PPP (current international $) (billions)	41.09	82.99	170.08	316.67
GNI per capita, PPP (current international $)	2390	4420	8230	14,290
People				
Income share held by lowest 20%	8.7	7	7.8	7.4
Life expectancy at birth, total (years)	72	70	73	76
Fertility rate, total (births per woman)	2.5	2.2	2.2	2
Adolescent fertility rate (births per 1000 women ages 15–19)	30	30	23	16
Contraceptive prevalence, any method (% of married women ages 15–49)	66	70	68	–
Births attended by skilled health staff (% of total)	94	96	99	–
Mortality rate, under-5 (per 1000 live births)	23	17	11	7
Prevalence of underweight, weight for age (% of children under 5)	29.1	22.9	21.6	–
Immunization, measles (% of children ages 12–23 months)	80	99	99	97
Primary completion rate, total (% of relevant age group)	97	107	99	98
School enrollment, primary (% gross)	109.9	107.8	99.5	100.3
School enrollment, secondary (% gross)	83	–	97	100
School enrollment, primary and secondary (gross), gender parity index (GPI)	1	–	1	1
Prevalence of HIV, total (% of population ages 15–49)	0.1	0.1	0.1	0.1
Environment				
Forest area (km^2) (thousands)	23.5	21.7	21	21.1
Terrestrial and marine protected areas (% of total territorial area)	–	–	–	3.4
Annual freshwater withdrawals, total (% of internal resources)	18.5	24.6	24.5	24.5
Urban population growth (annual %)	1.2	0.5	0.8	1.9
Energy use (kg of oil equivalent per capita)	321	443	471	–
CO_2 emissions (metric tons per capita)	0.22	0.58	0.63	1
Electric power consumption (kWh per capita)	153	295	451	–
Economy				
GDP (current US$) (billions)	8.03	16.33	58.64	88.5

(continued)

Table 7.1 (continued)

Country profile				
	1990	2000	2010	2021
GDP growth (annual %)	6.4	6	8	3.5
Inflation, GDP deflator (annual %)	20.1	7.3	26.9	8.5
Agriculture, forestry, and fishing, value added (% of GDP)	27	20	9	9
Industry (including construction), value added (% of GDP)	26	27	28	30
Exports of goods and services (% of GDP)	30	39	21	17
Imports of goods and services (% of GDP)	38	50	28	24
Gross capital formation (% of GDP)	22	28	24	37
Revenue, excluding grants (% of GDP)	21	16.8	12.3	8.3
Net lending (+) / net borrowing (−) (% of GDP)	−5.2	−8.4	−6.1	−11.7
States and markets				
Time required to start a business (days)	–	56	38	8
Domestic credit provided by financial sector (% of GDP)	–	–	–	–
Tax revenue (% of GDP)	19	14.5	10.9	7.4
Military expenditure (% of GDP)	2.3	5	2.7	1.9
Mobile cellular subscriptions (per 100 people)	0	2.3	84	141.3
Individuals using the internet (% of population)	0	0.6	3.9	66.7
High-technology exports (% of manufactured exports)	–	–	1	1
Statistical capacity score (overall average) (scale 0–100)	–	–	77	80
Global links				
Merchandise trade (% of GDP)	57	72	38	37
Net barter terms of trade index (2000 = 100)	82	100	110	117
External debt stocks, total (DOD, current US$) (millions)	5868	9250	21,684	56,592
Total debt service (% of exports of goods, services and primary income)	16.1	12.1	12.3	31.4
Net migration (thousands)	−36	−92	−27	−92
Personal remittances, received (current US$) (millions)	401	1154	4123	5522
Foreign direct investment, net inflows (BoP, current US$) (millions)	43	173	478	598
Net official development assistance received (current US$) (millions)	616	222.4	558.6	154

World Bank (2023)
Source: World development indicators database
Country: Sri Lanka
Data from database: WDI central
Last updated: 07/25/2023

> *Large garment manufacturers, again, a client of mine, they are very seriously looking at environmental discourses. Because even a factory being open, we are very much involved in that, to find out whether the factory creates any pollution. To that extent they are very sensitive. So before that, before they open the plant, they say, ERA assessment will be done. And you need to talk to the community. You have to talk to the opinion leaders in the community, and then buying from them, because otherwise they think that it's not the right thing to do. So I think that those initiatives are now happening.*

Participant SB3 also mentioned the impact of government corruption on sustainability. In the words of the participant:

Table 7.2 Major themes from Sri Lanka

Pillar of sustainability	Business leaders	Political leaders
Economic	Long-term stability	Long-term profitability
	Impact of globalization	
Social	Employment of women in a patriarchal society	Inequity in employment of women
	Lack of attention to discrimination against ethnic groups	Impact of globalization Government action
Environmental	Impact of businesses on resources	Influence of Buddhist philosophy
	Locals lacking environmental consciousness	Government action
		Experiences in a developing country

> *If you take a cross-section of very successful companies, they may have respective affiliations to various kinds of people, regimes of the country. Success is very strongly correlated to how strong the relationship is with such regimes. So I am questioning all … In the past 10 years, I am questioning, are there still any truly, truly entrepreneurs in this country?*

7.1.3 Employment of Women in a Patriarchal Society

Both business and political leaders interviewed believed that social sustainability was not achieved in Sri Lanka due to the employment inequity experienced by women. Business leaders insisted that the patriarchal society of Sri Lanka affected the employment of women. Participant SB3 stated:

> *I would say still it's a patriarchal society.*

Nonetheless, some participants believed that gradual changes in the employment of women were witnessed in recent years, as more women attained higher education. Participant SB1 shared:

> *I think you are going to the doctor's office now. It's full of women. There are very few men if you go to the universities.*

Participant SB3 stated:

> *But free education played to the equitable environment, so anybody has equal opportunity to get qualified and elevating their lifestyle. Then coming into gender inequality. I would say, now we are having 53% of women in Sri Lanka. Right? Roughly 53% women. And in the urban sector, there is relatively equal opportunity on meritocracy.*

Political leader SP1 believed that while more women were employed, women still experienced inequity in terms of labor rights. The participant shared:

> *Workplace, there are some good places that maintain ethics, and as a woman in the workplace and the rights we have as a female, mostly we have the breast feeding rights, we have bringing up children rights. Those are my constitutional rights, we have, and the… we are entitled to a good leaves system. And, maternity leave for women is good, but I have heard some work places, especially not ethically maintained, well, those places, they are not, not maintaining such things. But, I think generally, work force, females are in the labour sector, yes. Females are doing a lot of hard work. And I believe the women are more hard-working than men. And, of course, they are the breadwinners.*

7.1.4 Lack of Attention to Discrimination Against Ethnic Groups

Apart from women experiencing employment inequity, ethnic groups were also believed to be experiencing similar challenges. However, the business leaders believed that the issues experienced by ethnic groups lacked attention. Participant SB3 claimed that people generally lacked sensitivity to other ethnic groups, and reported:

> *Yes. It's a very sad story. I have been travelling and meeting people all over the country. As a nation Sri Lanka doesn't have these ethnic divisions by and large. There are some politically minded, or ethnically minded people. There are extreme values. So I'm talking about a median values kind of things. I worry Sri Lankans are not so sensitive unless otherwise woken up.*

What the participant was implying is that politicians provoke ethnic tension for their narrow political gain. This has led to civil war and lingering problems in a country struggling to progress.

7.1.5 Impact of Globalization on Social Sustainability

For political leaders, social sustainability was associated with globalization. Participant SP2 claimed that developed countries tend to be the source of inequality, and stated:

> *I think under this available system inequality is very common in the world because a lot of developed countries, they own a lot.*

Participant SP1 explained the inequality in terms of urban and rural areas. The more developed urban areas experienced comforts and luxuries, however, rural areas did not. In the words of Participant SP1:

> *Professor, having said that, this is very visible in the urban sector. But, when you go to the rural, the richness is not visible, right, because in the rural sector, people are sustainable. They grow their own food. They do their own work. And, monitoring their living is not as compulsory as urban sector. So, therefore, they made the system more sustained. But with the pressure of the development and the migration for the urban sector, this is becoming quite tight and that's ... because of that, the young generation is suffering. The pressure for them is too much, because youngsters at rural level, they don't want to remain in the village because they think the development is there in the city, therefore they have to come. And, therefore, once they come here, they are suffering. It's not sustainable, and they can't really survive here in the city. And, then they end up with lot of suffering and maybe getting drugs or lot of social issues.*

Participants SP1 and SP2 in the foregoing comments addressed two issues in the SDGs: inequality between developed and developing nations and the need for sustainable urban areas.

7.1.6 Government Action on Social Sustainability

To be socially sustainable, the political leaders generally believed in government action. In this regard, the Government may provide education subsidies, financial subsidies, and enact regulations. Participant SP3 shared:

> *Well, actually, Sri Lanka had a very strong socialist movement since 1930s. Because of that, they had a lot of subsidies in Sri Lanka. Education and health, total free of charge. From grade one to university.*

Participant SP2 believed that world leaders could do the same at the United Nations to sustain developing countries. The participant stated:

> *I think UN resolution is more important for us. Especially developing countries like us. But, they accepted, they mean that all the countries, all the policies accepted, but they can't implement. We haven't started. We can start all of these done in our countries because we don't have big factories, very huge, big factories. We cannot start our digital system. But America is very digital because they are huge land area. Started with small holders, we're going to start our digital with small holders.*

7.1.7 Impact of Businesses on Resources

Business leaders SB1 and SB2 perceived that Sri Lanka generally relied on the agriculture industry. They saw agricultural activities having an impact on the environment. Participant SB1 claimed that excessive use of chemicals on crops left residues on the land, which then affected livestock, and eventually, humans. Participant SB2 cited a possible link between environmental sustainability and climate change. The participant shared:

> *Climate change nobody can predict. We have never had this situation in the world 10-15 years ago. Winter is winter; summer is summer. Winter is being dragged on to be autumn also. I mean these*

are things that are happening or has happened. Of course, the activities have adverse effects. In agriculture I know this time of the year I have the water I plant. If the climate change comes and no water, how will the plant live? Now this last two weeks ago we had frost and the frost bite. All the leaves die.

7.1.8 Locals Lacking Environmental Consciousness

The business leaders generally perceived that locals tend to lack environmental consciousness and engage in consumerism. While Participant SB3 believed that locals were generally compliant with government regulations and were beginning to value the environment, participant SB1 believed that locals were affected by the globalizing world and tended to purchase and consume unnecessarily. Participant SB1 narrated:

I think we have to reeducate everybody. Not to focus on more and more consumption, more and more gadgets. You can live as comfortably, with a much simpler way, with far less consumption, and until we get to that, there is going to be a problem. And the rich countries don't want that to happen because they're dependent on us buying all their goods. Buy many cars. Buy more consumer products, but how are we going to? Particularly the world will never change. That's the problem. I don't think it's, well, it's really impossible.

7.1.9 Influence of Buddhist Philosophy

Three political leaders believed that the principles of Buddhism played a role in environmental sustainability. Participant SP2 explained that Buddhism was not just a religion, but a philosophy and a way of life. Therefore, incorporating Buddhist philosophy in lawmaking was generally accepted. Participant SP2 stated:

Spirituality, yes, spirituality. Our historical, cultural experiences and also spirituality, bring in all the things together we can find solutions. Especially with sustainable agriculture.

The participant further explained:

If the changes are in my mind, then one person can change by using; he can influence others. There's a way, there's a way, but if it's not enough under the capital influences, but your spirituality is more important. We need now a spiritual approach to the world. Last hundred years, influence by a political approach, it is not enough now.

Participant SP3 explained the relationship between Buddhist philosophy and sustainability similarly and stated that the concept of sustainability has long existed in Buddhism. The participant noted:

You are merely the custodian who poses on behalf of others. So you have to look of, you should ensure these resources will be used for everybody's benefit, including the trees and the birds. And you should further ensure forthcoming, you will utilize the resources in such a way that forthcoming generations will also be able to use in the very same manner that you use. Isn't it all about sustainability?

He continued:

Well, actually sustainability has been a novel concept for Western world. It's not something new to us. Being Buddhist, so Sri Lanka has been a country with Buddhist civilization, and Sri Lankans believe that Buddhism has been learned as a philosophy in a certain part of the world. And in other part of the world, they practice Buddhist rituals. But here we have transformed the Buddhist philosophy to fit a lifestyle. So most of the religions say that the world has been created by the God for the man. So man is in it, and he started to utilize the resource created for him; he destroyed the entire environment. But Buddhism says, "Man is only a part of the environment, which is interconnected, interwoven, and interdependent." So change in environment has a drastic reverse impact on the man.

7.1.10 Government Action

Concerning the previous theme, Sri Lankan political leaders also generally believed that nature was their "heritage," as described by Participant SP3. The participant emphasized that the government had the power to preserve their heritage instead of focusing on developing roads and buildings. The participant further argued that by claiming the natural resources, the people were given a sense of ownership in which people might feel obligated to protect what was theirs. Participant SP3 elaborated:

> *And you should protect it. Sometimes, you know? We're merely making them proud about what they have; we can motivate them to join hands in conservation of the environment. So as I said because of that religious aspect in our culture, also go, there are politicians, then government officers. But the true leader of the village is the Buddhist monk. Everybody respects him, irrespective of type of cultural heritage. Because of that, because of his leadership, people, as I told you before, it's a part of our culture, part of our heritage.*

The government was also believed to be responsible for leadership in protecting the environment. Participant SP1 reasoned that with the government applying "green leadership," environmental sustainability may be attained. The participant further stated that political leaders had the power to write evidence-based legislation concerning environmental sustainability. In the words of Participant SP1:

> *We talk evidence based, so the policy makers and all are lobbying through our research work and the existing data. So, we have strong connections with the policy makers or the government, so wherever we want, we speak up and we write letters. And, when we don't hesitate to write letters even to President level or whatever. And, then we have public gatherings, and while we're there, we want to raise our concern. We network with other organizations, so traditions and all these, that's all happening. But, other than that, we try to empower the public, so if the public got more enthusiastic on these issues with the knowledge and the facts and figures, and then they can, themselves, lobby and be more effective.*

7.1.11 Experiences in a Developing Country

Sri Lankan political leaders perceived that sustainability went hand-in-hand with development. As a developing country, the participants generally believed that Sri Lanka was at a disadvantage. Participant SP1 believed that the government had to choose to sacrifice one of the three pillars of sustainability due to lack of funds. Participant SP1 mentioned:

> *That's my biggest criticism. And, maybe outside of Sri Lanka it would be the same, but I know some case studies which is not true, anyway. I'll tell you, these three pillars in theory. Every person who talks about sustainability, they will draw nice circles, three circles and they say, "Yes, sustainability or sustainable development." And, but, if there is a road ... if we need to have the road here. What is the most sacrificing component, either people or the environment? Or the economy? It's the least concern and most targeted for the vulnerability, the component is environment.*

For Participant SP1, therefore, environmental concerns are secondary to economic and social demands.

Similarly, Participant SP2 claimed:

> *We have no space. We can't reverse. We can't implement. We can't start new things, new system. We can't implement it.*

The participant explained:

> *We need ... Locally we understand the situation, and we have to introduce some rules, some ethics to protect the environment and protect the people. I understood, last 20 years we already had so many global conventions under the UN, but unfortunately we couldn't implement anything. It was, you know, and me also know about that. Because capital ownership and all of the control of the capital, they can control all the policies. All the political leaders, whatever we proposed globally, internationally. But we can't implement anything.*

7.2 Summary of Themes from Interviews with Sri Lankan Leaders

Sri Lankan political and business leaders believed in the three pillars of sustainability. The themes that emerged from the Sri Lankan political and business leaders were:

1. Long-term impact
2. Impact of globalization on economic sustainability
3. Employment of women in a patriarchal society
4. Lack of attention to discrimination against ethnic groups
5. Impact of globalization on social sustainability
6. Government action on social sustainability
7. Impact of businesses on resources
8. Locals lacking environmental consciousness
9. Influence of Buddhist philosophy
10. Government action
11. Experiences in a developing country.

Sri Lankan political and business leaders interviewed believed in the long-term impact required for economic sustainability. While political leaders were focused on profitability, business leaders also believed in the stability of production. Business leaders also believed that the globalizing world affected economic sustainability, as developing countries attempted to catch up to developed countries. On social sustainability, both groups of leaders recognized the inequity experienced by women in employment. They reasoned that a patriarchal society hindered women from being employed; however, business leaders were beginning to see a shift in the workforce, as more women obtained higher education. Business leaders also pointed out the lack of attention given to the employment inequity experienced by ethnic groups. Political leaders again mentioned the impact of globalization, as local corporations competed in the global market to provide fair wages. Government action was perceived to influence social sustainability through appropriate regulations.

In environmental sustainability, private businesses' utilization of resources was believed to impact nature. It was felt that locals tend to lack environmental consciousness. However, political leaders argued that Buddhist philosophy guided the way of life of the citizens and influenced the law. The government was believed to be obligated to enact legislation to protect the environment and make people understand the stewardship they have towards nature. Political leaders also expressed the belief that being in a developing country created difficulties in achieving environmental sustainability. They were of the view that a lack of resources forces the government to sacrifice the environmental pillar of sustainability for Sri Lanka to be economically viable.

References

CIA (2020) The World Factbook: Sri Lanka. From https://www.cia.gov/library/publications/the-world-factbook/geos/in.html. Accessed 29 July 2020

CIA (2023) The World Factbook: Sri Lanka. From https://www.cia.gov/library/publications/the-world-factbook/geos/in.html. Accessed 20 July 2023

Sachs JD, Lafortune G, Fuller G, Drumm E (2023) Implementing the SDG stimulus. Sustainable Development report 2023. SDSN/Dublin University Press, Paris/Dublin. https://doi.org/10.25546/102924

World Bank (2023) Country profile Sri Lanka. From https://databank.worldbank.org/views/reports/reportwidget.aspx?Report_Name=CountryProfile&Id=b450fd57&tbar=y&dd=y&inf=n&zm=n&country=LKA. Accessed 31 July 2023

Case Study 5: Guyana

8

Abstract

This is the last of five chapters that analyze the interviews conducted with business and political leaders in selected countries. In this chapter interviews with leaders in Guyana are examined. The chapter begins with background information on the country's history, economy, people, and environment. This is followed by the identification and discussion of themes on sustainability that permeate the interviews. Themes that surfaced in the interviews with Guyanese leaders were long-term profitability, sustainability as a responsibility, adapting to change, employment equity and fairness, free education providing opportunities, politics and local culture, income inequity, private sector taking action, culture of locals as lacking environmental consciousness, and government taking action.

Guyana is the only English-speaking country in the South American continent. It is a former British colony and is both culturally and historically closer to the English-speaking Caribbean than to the rest of South America. It is in the northeast corner of South America and is bounded by the Atlantic Ocean to the North, Venezuela on the West, Brazil on the South, and Surinam on the East. Guyana covers an area of 215,000 km^2 and has a population of approximately 780,000; 90% of whom live on a coastal strip 16 km wide. According to CIA statistics, the literacy rate for this population is 88% (CIA 2023).

The original inhabitants of Guyana were Amerindians. When Britain colonized the country after replacing Dutch colonists, slaves were brought from Africa. After the abolition of slavery, some indentured laborers were brought from China and Madeira, but the vast majority were from India. The ethnic composition of the present Guyanese population reflects the country's immigration history. Today, Guyana's two largest ethnic groups are the descendants of the Indian indentured laborers and African slaves. According to the CIA (2019), the ethnic makeup of the Guyana population in 2012 was estimated as follows: East Indian 39.8%, Black (African) 29.3%, mixed 19.9%, Amerindian 10.5%, other 0.5% (includes Portuguese, Chinese, White). Guyana's emigration rate is among the highest in the world, manifested in 55% of its citizens living in foreign countries. The per capita GDP in 2021 was estimated by the CIA to be $21,900 purchasing power parity (PPP). The agency also estimated that in 1999, household income or consumption by percentage share, an indicator of inequality, was 1.3% for the lowest 10% and 33.8% for the highest 10% of Guyanese households. The country's Sustainable Development Goals index score is 67.4, placing it 96th of 166 countries assessed. It is on track to achieve 40.7% of its SDG targets

J. Singh, *Sustainability and the Sustainable Development Goals in Cultural Clusters*, Sustainable Development Goals Series, https://doi.org/10.1007/978-3-031-70618-9_8

but the percentage of targets on which there is worsening performance is 15.3%. Among the targets that are worsening are those related to Good Health and Well-Being (SDG 3) (Sachs et al. 2023). Further statistics on Guyana's people and economy are shown in Table 8.1.

The statistics shown in Table 8.1 should be kept in mind as the narratives of the respondents are read. This will help in making sense of their views on sustainability. As shown in the table, Guyana is a sparsely populated country of less than a million people of whom about 12% live on less than $2.15 per day (2011 Purchasing Power Parity). The per capita Gross National Income (GNI) was $US9410 and the Current Purchasing Power Parity (current international $) was $23,020 in 2021. In 2000, the income share of the poorest 20% of the population was only 4.2%. Another notable item from Table 8.1 is that life expectancy in Guyana had risen steadily from 62 years in 1990 to 66 years in 2021.

Overall, on environmental measures, the country fares well as almost three-quarters of its total area of 215,000 km^2 is forest that was minimally depleted between 1990 and 2021. The per capita CO_2 emissions in 2021 were 3.47 metric tons compared to the world average of 4.7 metric tons.

8.1 Narrative Themes from Guyana Leaders

This section contains the themes generated from the narratives of Guyanese leaders. Table 8.2 shows the overview of these findings. The subsections that follow contain the description of each theme, and the findings for Guyana leaders are summarized at the end of the section.

8.1.1 Long-Term Profitability

Both groups of participants, the business leaders, and the political leaders, were of the view that economic sustainability referred to long-term profitability. The business leaders generally believed that sustainability involved not just quick returns on investment, but the ability of a business to generate profit over time. Participant GB1, a banker for 40 years, claimed that businessmen were often not "looking for a quick buck." The participant emphasized the importance of adding value to the business to be able to maintain profitability over time. As such, the participant considered trust among the stakeholders as essential for sustainability. The participant also mentioned that loyalty was a valued trait in Guyana. Participant GB1 explained:

> *As far as banking is concerned, sustainability is the key because we operate on trust. That's the commodity we sell, trust. With that, you have to have strategy that ensures over the long-term that your bank will be there providing the services that persons need. Because in this country, people are pretty loyal, meaning to say, they are your long-term customers. So since that's the way they view you; you have to view yourself as well as being there over the long-term, sustaining a level of profitability.*

In addition, Participant GB2 asserted that business investors were looking to know where their money was going. Participant GB2 narrated:

> *Capital has to know where it's going, and the work of that capital has to be predictable. I must be able to have stable conditions in the next 10 years, if that is how long my project is gonna last. And if I can't do that, I can't invest my capital, so what I just described to you with Britain, is one such thing where that risk wasn't catered for and that person invested, invested money into plant exporting.*

For Participant GB4, the business' stakeholders included his employees. Taking care of his employees, "especially the top ones, the skilled ones," was needed for economic sustainability. Participant GB4 also believed that to be able to have a stable business, one must "keep out of politics" and get to know the market. The participant explained:

> *So I decided after a while, I knew the markets very well in the UK, Canada and I intended to go into Europe too, so that was done, and we started producing locally, first just for the local market. And then we decided, I had with me Ramesh Singh (pseudonym), who is now the Chairman of the company. I had sent him to the UK to get qualified,*

Table 8.1 Country profile: Guyana

Country profile	1990	2000	2010	2021
World view				
Population, total (millions)	0.75	0.76	0.75	0.8
Population growth (annual %)	−0.8	0.1	−0.4	0.9
Surface area (sq. km) (thousands)	215	215	215	215
Population density (people per sq. km of land area)	3.8	3.9	3.8	4
Poverty headcount ratio at national poverty lines (% of population)	–	–	–	–
Poverty headcount ratio at \$2.15 a day (2017 PPP) (% of population)	29.4	11.9	–	–
GNI, Atlas method (current US\$) (billions)	0.29	0.66	3.38	7.57
GNI per capita, Atlas method (current US\$)	390	870	4520	9410
GNI, PPP (current international \$) (billions)	2.02	4.37	7.02	18.52
GNI per capita, PPP (current international \$)	2700	5760	9390	23,020
People				
Income share held by lowest 20%	–	4.2	–	–
Life expectancy at birth, total (years)	62	64	67	66
Fertility rate, total (births per woman)	3.1	3	2.7	2.4
Adolescent fertility rate (births per 1000 women ages 15–19)	126	102	80	67
Contraceptive prevalence, any method (% of married women ages 15–49)	38	37	43	30
Births attended by skilled health staff (% of total)	–	83	89	98
Mortality rate, under-5 (per 1000 live births)	61	47	37	28
Prevalence of underweight, weight for age (% of children under 5)	13.2	11.9	11	9.4
Immunization, measles (% of children ages 12–23 months)	73	86	95	94
Primary completion rate, total (% of relevant age group)	91	108	93	–
School enrollment, primary (% gross)	96.7	105.5	96.5	–
School enrollment, secondary (% gross)	77	102	90	–
School enrollment, primary and secondary (gross), gender parity index (GPI)	1	1	1	–
Prevalence of HIV, total (% of population ages 15–49)	0.1	0.7	1.3	1.4
Environment				
Forest area (sq. km) (thousands)	186	185.6	185.2	184.2
Terrestrial and marine protected areas (% of total territorial area)	–	–	–	5.2
Annual freshwater withdrawals, total (% of internal resources)	0.6	0.7	0.6	0.6
Urban population growth (annual %)	−1.1	−0.2	−1.3	1.3
Energy use (kg of oil equivalent per capita)	511	–	661	–
CO_2 emissions (metric tons per capita)	1.57	2.19	2.35	3.47
Electric power consumption (kWh per capita)	–	–	–	–
Economy				
GDP (current US\$) (billions)	0.4	0.71	3.43	8.04
GDP growth (annual %)	−3.1	−1.4	4.1	20.1
Inflation, GDP deflator (annual %)	56.4	6.6	4	22.5
Agriculture, forestry, and fishing, value added (% of GDP)	38	30	29	13
Industry (including construction), value added (% of GDP)	17	20	25	50
Exports of goods and services (% of GDP)	63	96	–	–
Imports of goods and services (% of GDP)	80	111	–	–
Gross capital formation (% of GDP)	31	24	–	–

(continued)

Table 8.1 (continued)

Country profile	1990	2000	2010	2021
Revenue, excluding grants (% of GDP)	–	–	–	–
Net lending (+) / net borrowing (-) (% of GDP)	–	–	–	–
States and markets				
Time required to start a business (days)	–	–	23	18
Domestic credit provided by financial sector (% of GDP)	265.9	100.5	32.2	50.9
Tax revenue (% of GDP)	–	–	–	–
Military expenditure (% of GDP)	0.9	1.8	1.4	1
Mobile cellular subscriptions (per 100 people)	0	5.2	74.9	106.4
Individuals using the Internet (% of population)	0	0.6	29.9	84.8
High-technology exports (% of manufactured exports)	–	–	0	1
Statistical capacity score (overall average) (scale 0–100)	–	–	53	46
Global links				
Merchandise trade (% of GDP)	143	150	66	109
Net barter terms of trade index (2000 = 100)	–	100	127	155
External debt stocks, total (DOD, current US$) (millions)	1963	1374	1408	1890
Total debt service (% of exports of goods, services and primary income)	18.7	10.2	2.7	6.6
Net migration (thousands)	−22	−13	−14	−4
Personal remittances, received (current US$) (millions)	1	27	368	548
Foreign direct investment, net inflows (BoP, current US$) (millions)	8	67	198	2237
Net official development assistance received (current US$) (millions)	164.8	63.6	106.4	138.8

World Bank (2023)

Table 8.2 Major themes from Guyana

Pillar of sustainability	Business leaders	Political leaders
Economic	Long-term profitability	Long-term profitability
	Sustainability as a responsibility	Adapting to change
Social	Employment equity and fairness	Politics and local culture
	Free education provides opportunities	Income inequity is an issue
Environmental	Private sector is taking action	Government action despite costs
	Culture of locals as lacking environmental consciousness	
	The government is leading the movement towards environmental consciousness	

> *come back, and I discovered he was a good young man to be trained. So I brought him into the fold, and Ramesh and I, we decided we gotta change the UK, Canada and whatnot. We gotta go into brands, that's where the money is, but you have to have the right brand, and we started preparing our product.*

Participant GB3 stated that economic sustainability referred to "something lasting into the future." The participant added that a sustainable business continuously grows. Participant GB3 emphasized the importance of resources, particularly renewable resources, to sustainability. In the words of the participant:

> *So the economic sustainability, of course the focus is on the renewable resources, as opposed to resources that can be easily depleted would allow for business models that can stand the test of time because, in the case of the extractive industries, when those resources have been depleted, that becomes the end of that particular industry.*

Participant GB2 was of the view that long-term profitability required long-term planning. Without a quick return on investment, participants GB1 and GB4 mentioned that businesses involving sustainability may not be a popular option. Participant GB1 stated:

> *This may involve decisions that are not necessarily popular with everyone, so that's where ethics comes in. If one is gaining the sustainability, one has to understand that not all decisions will be popular with everyone. Still, there will be some people who will be looking for quick returns. I don't blame an investor who wants quick returns.*

The participants who were politicians also believed that sustainability involved "continuity and probability," as stated by Participant GP2. Three of the five politicians—participants GP1, GP3, and GP4—believed that providing goods and services well was a sign of sustainability. Participant GP1 stated:

> *I would say, like a famous management scholar, that profits are not the endpoint, but profits are a measure and an indication of whether the goods and services are being provided well. But every enterprise needs to pay its way, if not, it's not gonna last. Of course, you have storages and accumulations from the past.*

Participant GP4 highlighted the importance of having well-trained staff and functional equipment. Participant GP3 claimed that being able to provide services despite a low budget meant that Guyana was "sustainable." Participant GP3 explained:

> *They have to do with the goals which we have set. And if we are able to fulfill those goals in a way that we can sustain within the national purse. Guyana is a poor country. I would have said if you go back into history—a mis-developed country for many, many years, both in colonial time and post-colonial times. The national budget is in fact the reservoir from which you would decide how much you can do and for how long. And it is in that context the issue of sustainability comes in. So in that sense, it shows the limitation of resources, and in fact from a sociological point of view, sustainability has to do with accessing scarce resources. Whether those resources are land resources, they are social resources, health, whether it's education, whether it is housing, and the other social affairs. Then you have to deal with the scarce resources of revenue to be able to impact on those.*

Participant GP3 also mentioned that the Guyanese government was guided by realistic goals and actions and followed the 17 sustainable goals set by the United Nations. Participant GP3 stated:

> *Well, they now have these new set of 17 sustainable goals, and we are committed to all of them. And they are all interrelated. They are all dialectically interrelated in the sense that you couldn't focus on one without focusing on the other. If you focus on poverty reduction, you have to deal with the issue of income generation and the sustainability of dealing with poverty reduction. What are the measures you're gonna be taking for poverty reduction?*

It should be noted that Participant GP3 is a senior government official and veteran politician. He was one of a small number of interviewees in the study who were aware of the United Nations Sustainable Development Goals.

8.1.2 Sustainability as a Responsibility

This theme emerged from the interviews with Guyanese business leaders. The business leaders generally felt that sustainability was the responsibility of businesspersons. Participant GB1 believed that this responsibility involved "good corporate governance." For participant GB1, an economically sustainable business had a leader who made good decisions, was concerned about the investors—not the competition, anticipated highs and lows in business, and knew when to stop. Participant GB1 added:

> *Many a time you have seen conservative managers look like fools, and you have to come to your board, and the board says, "But you're not doing enough because look at these guys over there, and they get huge profits." You might lose your job. Because time is not on your side. Later on, you may go back to your board and say, "See what I'm telling you. Look at those guys, they having trouble." So to put it... from the last point I made, good corporate governance has to be a part of sustainability, or it's got to have the courage to make the right decisions on a long-term basis. That takes some courage.*

Participant GB4 shared his experience as the head of his business, negotiating with different suppliers and the government to ensure the sustainability of the business. Participant GB4 revealed:

> *I negotiated with sugar companies, I negotiated for tractors, combines, and all these things in my company on the long term, and I went off to the UK, then to the US to bring these things, so my boss said, "You can do it. You know this company, and therefore you'll know the giant corporation. You're a member of the Sugar Producer's Association, and everything is merged into one for the locations in the Caribbean and elsewhere."*

According to participant GB2, businesses have an "obligation" to the environment, especially when using natural resources. Participant GB2 articulated:

> *We have an obligation to ensure we maintain the environment, the natural environment in its pristine state. We have an obligation to establish rules in business that will ensure that we do that. We also have an obligation to ensure that we practice good corporate social responsibility in the societies and communities in which we exist. So Guyana is involved in sustainable harvesting of the forest. We are also involved in sustainable mining of minerals. There are conditions now where you have to mine for your gold and your diamonds, and you have to put back the environment into its original state as much as possible.*

8.1.3 Adapting to Change

This theme emerged from the interviews with political leaders. For the majority of political leaders, sustainability involves the ability to adapt to changes. Some participants discussed the impact that the discovery of oil in Guyana might have on local businesses. Guyana's per capita oil reserves are one of the highest in the world with some 11 billion barrels estimated for a population of around 740,000. Participant GP4 stated that while oil may be a source of income in the present, climate change may lower the people's "consumption of fossil fuels." Participant GP4 explained:

> *Well, in Guyana, it's not so much an impact now as we speak, but of course Guyana has just discovered oil. If the struggle against climate change proceeds, it will affect the consumption, reduce the consumption of fossil fuels, and therefore Guyana stands to lose somewhat.*

Participant GP1 addressed social sustainability in relation to economic sustainability and employment equity. The participant claimed that businesses evolve depending on the trends and needs and provided the evolution of Dutch State Mines as an example. The participant shared:

> *Then [100 years ago], they were a coal company. [In the] 1950s, they were a chemicals company, and now in 2015/16, they are life products company.*

The participant drew a parallel to economic and social developments in Guyana. In this regard, the participant stated:

> *I would say that our way of living is evolving all the time, in totality. There's the individual responsibilities and where we are individually, but I think in time to our economic models, I believe we need to add two more constraints. One is for employment fairness, whatever way we might think of it. And two might be, another line would have to do with a certain amount of… a limit on inequality of incomes.*

Participant GP3 claimed that free education, availability of natural resources, and building infrastructure have been used to help Guyana adapt to economic changes. In adapting to changes, the participant claimed that the government was looking for new means of income. For instance, in education, students who received free primary and secondary education were beginning to pursue tertiary education, which was not free. Therefore, student loans were introduced. For natural resources, the participant provided potable water as an example, where access to clean water was once free. Later, the government charged for water due to "cost recovery." The participant added:

> *while the state could do the infrastructural work, people have to pay for water in a sense, even though it is highly subsidized.*

Further, Participant GP3 mentioned that the Guyanese economy used to rely on rice, sugar, and bauxite, and now oil and gold are being mined. Participant GP3 then stated:

> *Gold is now perking up our economy. We are likely to have even for the first time unprecedented so over 600,000 ounces of gold in one year, at a time when the price of gold is now trending up again. But the question is: Is that sustainable? If tomorrow the price of gold falls, and if the fuel import bill doesn't remain the same because now you have a fall in the price of imported fuel, and we depend 100% almost on fossil fuel for our electricity supply. The price has gone down; it is not where it used to be—$150 per barrel. It is now in the $30 or $40 per barrel. Now if the price of oil goes up, everything else would go up. So our question of sustainability has to do with the interdependent and interrelated trade relations with the rest of the world. So it's not like your John Donne concept of no man is an island; well it is like a John Donne concept no man is an island entirely by himself. We're all interconnected.*

8.1.4 Employment Equity and Fairness

This theme emerged from the interviews with the business leaders. Most of the participants perceived that there was a lack of social sustainability in terms of equity and fairness in employment, especially in gender and race. Participant GB2 expressed the opinion that women tend to experience discrimination. Participant GB2 shared:

> *Well, that's just gender first. Gender, we're doing very badly there; the corporate world is dominated by men. I don't think we treat female professionals with respect. I believe we have a long way to go in that respect and regard. I believe women need to stand up for themselves a little bit more and be sure that their voice is heard. With women's lib and all that, you know, I don't. That sparks a negative in the male and probably works to the disadvantage of women in societies like ours, but I think a lot needs to be done.*

Participant GB2 also claimed that people tend to prefer to work with people of the same race. The participant stated:

> *That's the thing, nobody would openly do it, because imagine I go to my HR executive or manager and say, Look, let's employ 50 Indians and 50 Afro-Guyanese. It's a difficult thing to tell that person because he or she should be employing the best person for the job. Unfortunately, deep down in our minds, we have an inbuilt bias mechanism that works.*

He continued:

> *In terms of race, there's a natural tendency whenever a company is run by a person of a particular race persons of the same race gravitate towards that company. You know, it's a natural phenomenon. But I think real efforts could be made in the corporate world, could be a lot of balance. I think balance is needed. We did consciously adhere to it; subconsciously I don't know. I am very conscious of ethnicity, and everything I do I try to balance. I try to use diversity principles and diverse teams, and I have an obligation to younger people to make them appreciate each other; appreciate each other's jobs and that kinda thing.*

The participant continued by saying that the cultural norms of the Guyanese may also influence social sustainability and employment equity. Participant GB2 cited that with the beginning of the oil industry, several jobs opened; however, the culture of "laziness" prevented people from getting or keeping jobs. Participant GB2 elaborated as follows:

> *Oil may do something, but I don't believe in handouts. I believe everyone should work hard with their hand to the wheel, and we have a lot of laziness in our society too. People tell you that they can't find a job; you give them a job, and they don't wanna work.*

Nonetheless, Participant GB2 explained that the government was beginning to act on social sustainability and started making agreements with the oil industry to open more jobs and hire more locals. In addition, Participant GB3 opined that businesses were also beginning to hire people based on their skills rather than other factors. Participant GB3 stated:

> *We still haven't gotten to the stage where our systems are as objective as they should be; therefore, judgment still slips in, and judgment does unfortunately get affected by issues of gender, race, other background information. However, there is a trend towards appraising and recruiting individuals based on their competence versus other issues which, you know, really make it very subjective.*

Participant GB4 stated that he has issued public shares for his business to help people generate income. Furthermore, the participant perceived that social sustainability applied to his staff and claimed to treat his employees well to retain them. Participant GB4 shared:

> *The ability to continue depends on your employees. You don't treat your employees well, they're not going to give you the productivity. Always treat your employees as first and foremost. Of course, they have to do their jobs. Once you do that, you get their confidence, the rest I, in my opinion, follows. So sir, that's my story.*

While issues such as limited budgets and government discrimination in employment (as pointed out by participants GB1 and GB3) were discussed Participant GB1 revealed that his business did not tolerate income inequity. The participant stated that in his company employees are paid a fair salary based on merit "Because, of course in banking, it's important to have persons who are financially stable." Participant GB1 noted:

> So then there is no gender discrimination in regards to pay. We pay above the minimum wage, way above the minimum wage. We set our wages ourselves, as far as yearly increments are concerned. We try to benchmark every year to the rate of inflation. So we try to do what is called a COLA, cost of living adjustments.

Participant GB4 was of a similar view:

> *You see, if you don't treat your employees well, pay them well, and also ensure they get benefits, you're not gonna get the best from them. And they could even sabotage the business.*

8.1.5 Free Education Provides Opportunities

This theme emerged from interviews with both business and political leaders. Some participants believed that social sustainability involved creating opportunities and that free basic education provided those opportunities. In this sense, education was not limited to primary, secondary, and tertiary schooling, but also includes continuous learning in the workplace.

Participant GB2 believed that providing education may present a "way out" for Guyanese. Participant GB2 mentioned:

> *Well you know the president is very strong on this. He sees that education is the way out. We have a lot of problems here in Guyana. So if you're poor, you're gonna be poor for the rest of your life because you can't afford to send your children to primary school, so they don't get an education although it's free. But as long as you don't get an education, you'll be underpaid, you'll be abused, and you will be used.*

While participant GB3 acknowledged that education did provide opportunities for the locals, he felt certain classes and people who had family businesses tended to have more opportunities to earn higher incomes. Participant GB3 narrated:

> *Education and the level of education achieved would usually determine prospects for employment. Lower level of education usually limits the ability to earn, however, we also have, to some extent, classes which exist in our society. The business class where family-owned businesses are still very prevalent and can, by virtue of belonging to such a family-owned business, generate revenues for participants in that field, whereas those who do not belong to those classes don't have that particular opportunity.*

8.1.6 Politics and Local Culture

For the Guyanese political leaders, social sustainability involves politics and the local culture. Participants GP1, GP4, and GP5 all discussed the impact of politics and political stability on social sustainability, particularly in employment equity and income inequity. Participant GP4 expressed the view that an imbalance of power in society may lead to instability, which may affect employment equity. Participant GP4 explained:

> *Look what happened in the United States. On the wider disparity of wealth you get the Donald Trump, you get a Brexit, then you get to Donald Trump. Now there's an election in France, Marie Le Pen might potentially win the presidency. You*

also have governments in Hungary and Poland which are extremely right wing. That is the effect of the ... the inequality in the United States will increase with reduction of taxes by the Trump administration. So it will get worse, not better. It creates social and political instability. It brings into office the right-wing ideologues, neo-fascists even. So it's very dangerous (to) political stability.

Similarly, Participant GP5 articulated:

Unemployment is a problem at the moment. I think the major problem in Guyana is security. Public security, the crime rate as you see, is spiralling and that can affect so many other things. Because of that some people with skills, and the brain drain, people will go away because of that, businessmen. I'm relating to your question about businesses, it can affect the businesses of a lot of people, because they will close the business, as it has been happening. A lot of people are afraid to open their businesses and so on.

Nonetheless, Participant GP1 believed that employment equity, especially for women, was becoming less of an issue, as more businesses were hiring women. However, in terms of race, the participant believed that people were still culturally/ethnically biased. Participant GP1 stated:

I think to some extent that is so, but I think human beings also have strong ethnic sensitivities. It's not genetic but socialization tends to perpetuate it. I say that because if a person grows up in one group ... I know cases both ways. In Guyana, we've Indians who grew up in Indian areas, even experienced the troubles in '64, and they tell me they converted. There's some of that the other way, too. Ethnic differences, which is not only physiological, but also other ways, the ways you make jokes are different in the groups still. All those things are strong compared with class.

In addition, Participant GP1 perceived that population density may have deterred development in general. The participant shared:

In Guyana, we have a lot of issues with development. I've been thinking more recently that our population size and population density are probably great limitations. If you were to have the same population density as in Trinidad, we should have 60 million people in our 83,000 square miles. Or we should limit our three-quarter million people, all of us, to live in 1,500 square miles. Population density is an important factor for development and sustainability. I guess it gets a little bit into economic sustainability and development and so on.

8.1.7 Income Inequity Is an Issue

While the business leaders believed that income inequity was becoming less of an issue in social sustainability, the political leaders argued that for a developing country such as Guyana, income inequity was still an obstacle. Although employment opportunities have been available locally, Guyanese have also been demanding higher pay. In addition, with access to education, the political leaders generally believed that Guyanese desired to be paid as much as they would be paid for the same job when working in a developed country. Participant GP1 elaborated on this as follows:

I think that's a difficult issue. I would say it tends to be worse in developing countries because there're a number of locals who can ... impose may not be the right word, but who are able to garner for themselves a rate of fee that is similar to abroad. Similar to what they would get in North America.

Participant GP1 further expressed the view that even when international aid was provided by organizations such as the United Nations, a Guyanese national's pay was not commensurate with that of a person from a developed country doing the same work. In the words of Participant GP1:

Actually, to a lot of the programs of assistance, I guess particularly in small countries like Guyana, are tending to distort the situation even worse. At times, maybe not large in number, say UN or an aid program, might say, "We want somebody for this job with these particular qualifications." And then even if it's a Guyanese, they hold out for US\$10,000, \$15,000, \$20,000 a month, in Guyana.

In a similar vein, Participant GP4 stated that once Guyanese acquired certain skills, they preferred working overseas. Participant GP4 perceived that competition in the global market was not socially sustainable. Participant GP4 stated:

Governments need to encourage investment in areas where the businesses can be sustainable, particularly in relation to ... in Guyana, at least in IT business in that aspect of service, Guyana is well-positioned to enter the outsourcing industry. As yet there is no big incentives for people to get involved in those aspects of industry and our skills leave very soon after being qualified—just disappear out of Guyana. There are great challenges to

> *maintain that sustainability or nurses go away or teachers go away or students go away mainly to Canada, where they get admission very easily. It's a very wide issue, but those are the challenges that countries like Guyana face.*

Participant GP2 believed that some businesses merely comply with the minimum wage requirements and government-mandated benefits; however, the government was encouraging businesses to raise pay as an act of "corporate social responsibility." Participant GP2 shared:

> *OK, well, we are guided by the labour laws and by the income tax Act and so on. So there is a minimum wage that, by law, you're forced to pay. We will try to encourage companies to pay over and above that, and especially at Christmas time and so on, to offer as a bonus if they have made profits. Some comply, some don't comply, because they would say, well, we have met the minimum requirements which explain minimum salary. But we also try to encourage corporate social responsibility.*

One participant, GP5, believed that the gap between the rich and the poor was closing in Guyana. Participant GP5 reported:

> *Yes, I think so, the inequality, the gap between the rich and the poor. Even though some analysts say this, the poverty gap has been reduced over the last 20 to 25 years. But there are still pockets of inequalities existing between groups and classes in Guyana. But I think, my own view is that it has been reduced quite a bit, the poverty gap. I remember there was a figure previously, of 85% or 86%, people who were below the poverty line. With the previous government, it came down to about 30% or 35%. I think the gap is reducing, but there are disparities.*

The disparity to which Participant GP5 referred was the disparity between income and wealth of those at the top and those at the bottom. Hence, while actual poverty levels have fallen, the gap between the rich and poor has not diminished. This is a phenomenon that was observed not only in Guyana but in the other four countries covered in this study.

8.1.8 Private Sector Is Taking Action

The business leaders generally believed that the private sector was acting towards environmental sustainability and cited specific instances. Participants GB1 and GB3 claimed to work in a "green business," which involved activities such as recycling water and using solar power. Participant GB2 shared:

> *I would say that, you know, Guyana for a small country with a small corporate population, companies here are really conscious of what is required in terms of social responsibility, environmental responsibility, and of course the economics. We are committed to pushing our companies in the right direction when it comes to those things. I could tell you, I work for this company where we recycle our water; to recycle our water the environmental cost of our plants is enormous. Every piece of machinery that is bought, is bought for the environment.*

Participant GB3 mentioned having an office that was "fully solar-powered." In addition, Participant GB3 believed that to be environmentally sustainable, one must treat natural resources as assets to be taken care of. The participant claimed:

> *Our practices allow us to generate revenues for the country. However the resource we use is regenerated. Takes a 30 to 50 year cycle, however, that is the possibility. We then have to carefully weigh where we use such areas from mining as well. We sometimes ... the areas are completely stripped of all vegetation, which sometimes permanently damages the ability of the soil. The resource is removed, and then also the soil is done. If no significant reclamation projects are put in place through day-to-day regeneration, I think for the past decade or so we have seen a move to utilize our resources in a more responsible manner, so that is very heartening. However, state planning to an extent that it ... objective new ways or assets. That's what they really are, our assets. And then to make the best use of those assets is a move that in Guyana can ensure that we have resources to support our social standards for the long term.*

Participant GB2 mentioned being familiar with Participant GB3's solar-powered office and noted that there were several environmental advocates in Guyana. Participant GB2 reported:

> *We have an environmental advocate in the city, on our council, in the private sector commission, and we have become very conscious of the environment. As I speak, this evening a company [bank] is launching a totally solar-powered bank.*

Participant GB1 claimed to be helping build a "green economy," and that big businesses helped small businesses start "green" projects through "moral persuasion." Participant GB1 shared:

> *Long before this government in there started promoting a green economy, we have thought how we should try to promote projects that are friendly to the ecosystem. To do that, we give a preferential rate of interest. We try to get the government to buy into it, and subsidize that interest rate by giving to us a tax break. But we haven't succeeded as yet. But it's there with the lower interest rate than market rate; it's for anything ... and it can be very wide. It's for anything that reduces the carbon footprint while engaging in some productive activity.*

8.1.9 Culture of Locals as Lacking Environmental Consciousness

The business leaders interviewed were of the opinion that the general population of Guyanese tended to lack environmental consciousness. Participant GB1 asserted that "misinformed and/or uneducated people" tended to take natural resources for granted and treated natural resources as business opportunities. The participant stated:

> *We have always taken our natural resources for granted. We are a small population, with a huge land area. Take forestry for instance. We hear some people that are saying we are cutting down too many of our trees.*

Participant GB1 added:

> *In terms of protecting our part of the planet, I don't think we are very good. I don't think we feel responsible enough. But that's not such a bad thing for people who are not widely read on climate change. I hear Mr. Trump says climate change is a hoax. I hear he's going to put back in his coal mines.*

The participant briefly mentioned Guyana's agreement with Norway regarding being paid for carbon credits and expressed the view that the natural resources of Guyana were treated as commodities for trade. The participant narrated:

> *We have an arrangement with Norway, as you know, carbon credits, because they want us to not cut down any of our trees. So they offered us US$250 million, which some people thought was a paltry sum for not cutting some of our forest down or keeping it at a certain level. They've established the levels at which we should go off. Otherwise they wouldn't be ... I think it's five years; they will give us the $250 million over a period of five years. I think we've done about three or four years already.*

Participant GB1 argued that "poor people" tend to think about survival over sustainability. The participant reasoned:

> *Poor people don't have time for sustainability. I mean, take Haiti for instance. They have cut down all their trees to make firewood, to exist. So now Haiti is bare. I am told if you fly over Haiti and the Dominican Republic, you can see where Haiti stops and the Dominican Republic begins. One side has trees, and the other side does not. So poor people, sustainability is a luxury for them. They don't even think about.*

Participant GB3 expressed a similar view and cited Maslow's hierarchy of needs, in which people from a developing country like Guyana prioritized the satisfaction of lower-level needs such as food, shelter, and security over sustainability. Participant GB3 emphasized:

> *Looking at it from ... through the lenses of Maslow, it's a fourth- and fifth-level type of motivator for sustainability, and unfortunately, the majority of our population is still at anywhere between level one to three. So it's not a very high priority on their agenda. So, while we know that this is the direction that we need to head into ... We also understand that the majority of our population has other things on their minds.*

8.1.10 The Government Taking Action

The interviewed political leaders contended that the government was acting towards planet sustainability. According to participants GP3 and GP4, the government was moving towards a "green economy." Participant GP3 shared:

> *So our sustainable development has to do with looking at a green economy. Looking at the possibility of building a green state, where you utilize your internal natural resources to produce basically energy, which is perhaps taking the chunk,*

the largest single chunk of your financial allocation, or your financial reserves. So we can become sustainable to the extent that we are able to harness and to utilize the new resources that will bring down the cost of production, the cost of manufacturing.

Participants GP2 and GP4 claimed that the government was focused on policymaking to enact laws on environmental sustainability. Participant GP2 reported that businesses needed environmental and social impact assessments before operating. The participant shared:

So there is more consciousness, more awareness. And they put pressure on the business arm to ensure that they comply. In the past, people put in a gas station or some big multi-farm without any environmental and social impact assessment, so they were using chemicals and so on. They just dumped it, polluted water, polluted farmlands, and so on. Now, that can't happen. You have to do an environmental and social impact assessment. The consultant has to register with the environmental protection agency. Once he or she is registered, then you can choose that person. And there are very strict guidelines. The public has an opportunity to comment on it as well.

Participant GP4 disclosed that the government has implemented some environmental policies, for example, limiting the use of mercury in Guyana. The participant stated:

There has been an effort by the government of Guyana, the past government and the current government, to reduce the use of mercury. Sustainability is very important in relation to both the environment and business, and one needs to encourage investment. Governments need to encourage investment in areas where the businesses can be sustainable particularly in relation to … in Guyana at least in IT business in that aspect of service Guyana is well-positioned to enter the outsourcing industry.

Participant GP2 expressed the view that the policies made by the government for environmental sustainability were influenced by the public. Participant GP2 elaborated as follows:

Their actions are being watched by the general public who take our responsibility very seriously. So even though the businessmen or business sector might try to water down these things, I think if you have a responsible government, if you have responsible public, and a responsible course, that they will try to ensure that the balance is struck.

Participants GP3, GP4, and GP5 all noted the agreement between Guyanese government and the Government of Norway. Participant GP5 mentioned:

The Norwegian agreement, where we are paid for this sort of thing, where we are being rewarded for it.

Participant GP4 shared:

But hopefully, you know the government has been arguing, the past government and the present government, have been arguing that Guyana should be paid for the sustaining of our forests, which contributes. Hopefully those things will come on stream at some time—that's with the Norwegian agreement. If payment for forest comes on stream and is recognized by the rest of the world. It is now recognized by Norway but nobody else recognizes it.

8.2 Summary of Themes from Interviews with Guyanese Leaders

The major themes generated from the Guyana leaders' interviews were:

1. Long-term profitability
2. Sustainability as a responsibility
3. Adapting to change
4. Employment equity and fairness
5. Free education provides opportunities
6. Politics and local culture
7. Income inequity is an issue
8. Private sector is taking action
9. Culture of locals as lacking environmental consciousness
10. Government taking action

Both groups of participants—the business leaders and the political leaders—believed that long-term profitability was involved in economic sustainability. In addition, both groups of participants perceived that the government was acting towards environmental sustainability.

Long-term profitability, for business leaders, involves generating income over some time rather than making quick returns on investments. The participants believed that stability must be established for the country to achieve economic sustainability. Trust, planning, and adapting to changes were perceived to be essential in business stability. In addition, business leaders believed sustainability was a responsibility: people in business needed to make good decisions and practice good corporate governance. For political leaders, being able to provide goods and services well was a sign of economic sustainability and economic sustainability also included adapting to changes and the ability to look for new means to generate income.

For social sustainability, business leaders emphasized opportunities. The participants perceived that more diversity was needed in the workforce. Employment equity and fairness could be improved through government regulations, hiring more women, and changing the "laziness" mindset of locals. Furthermore, the business leaders also believed that free basic education improved social sustainability through opening more opportunities. As a result, people were being appropriately compensated and income inequity was becoming less of an issue. For political leaders, on the other hand, income inequity remained an issue. The political participants explained that as more people gained education and acquired skills, the more qualified they were for jobs. As such, people desired to be paid higher wages. Politics and the local culture were also perceived by the political leaders to influence social sustainability, political stability, and cultural sensitivity may have an influence on employment equity and income.

Environmental sustainability was seen as being significant by business leaders. The private sector was believed to be taking action toward a "green economy." Businesses were using renewable energy and helping start-up businesses to begin using sustainable materials and resources. However, the Guyanese business leaders interviewed believed that the culture of locals did not include environmental consciousness. Being residents of a developing country, the Guyanese were generally focused on income and did not prioritize the use of sustainable materials due to costs. Both groups of participants believed that the government was taking action to achieve environmental sustainability. Some participants mentioned the government's agreement with the Norwegian government for funds in exchange for carbon credits. The political leaders emphasized the significance of policymaking on environmental sustainability stating that, among other things, impact assessment before operations begin increased the sustainability of a new business.

References

CIA (2019) The World Factbook: Guyana. Accessed 5 Feb 2019 from https://www.cia.gov/library/publications/the-world-factbook/geos/gy.html

CIA (2023) The World Factbook: Guyana. Accessed 20 July 2023 from https://www.cia.gov/library/publications/the-world-factbook/geos/gy.html

Sachs JD, Lafortune G, Fuller G, Drumm E (2023) Implementing the SDG stimulus. Sustainable development report 2023. Paris: SDSN, Dublin: Dublin University Press. https://doi.org/10.25546/102924

World Bank (2023) Country profile Guyana. Accessed 30 July 2023 from https://databank.worldbank.org/views/reports/reportwidget.aspx?Report_Name=CountryProfile&Id=b450fd57&tbar=y&dd=y&inf=n&zm=n&country=GUY

9 Inter-role and Inter-cultural Analysis

Abstract

Political leaders and business leaders see the world differently. The former are more likely to be ideologically oriented and the latter more pragmatic. Further, national culture has been shown to influence beliefs and actions. Sri Lanka and India are part of a cultural cluster while Canada and the USA are part of another. In this chapter, the interviews conducted in the two clusters were analyzed to identify similarities and differences in leaders' views on sustainability according to culture and role. These are presented graphically and selected themes are considered in greater detail.

As previously mentioned, and is widely believed, political leaders and leaders of institutions view the world differently. Singh (1984) found a marked difference between the views of education leaders and political leaders in Guyana on the role of schools in national development. This difference is essentially a result of the ideological orientation of political leaders versus the supposed pragmatic orientation of business and other nonpolitical leaders. Bayley and Kayes (2018) in an article in *Psychology Today* discuss the various ways in which business and political leaders differ. The election of Donald Trump as President of the United States illustrated the appeal of seemingly successful businesspeople to the electorate. However, the apparent failure of Trump as a policymaker may signal caution in assuming a transference of skills from the business to the political world. Several other businesspeople such as Oprah Winfrey, Mark Cuban, and Howard Schultz have apparently recognized this and opted not to seek political office.

Bayley and Kayes (2018) suggest three reasons why business leaders may not be suited for the political world. First, they suggest that business leaders may not be comfortable in an environment of checks and balances which is not the norm in the business world where leaders could pursue efficiency without such restrictions. Second, they posit that corporate leaders would be challenged to balance the competing interests of a myriad of stakeholders that are characteristic of political policymaking, as business policymaking involves a limited number of stakeholders. According to Bayley and Kayes (2018), the level of risk that corporate leaders take simply cannot be taken in public policymaking.

Risk-taking may be a useful attribute for businesspeople, but in politics it could be a liability. Perhaps someday the handling of the COVID-19 pandemic by President Trump will be used to illustrate the danger of having a risk-taking businessperson oversee the health care of a nation. Bayley and Kayes (2018) concluded that: "Politics, at its heart, requires building and maintaining a coalition of support. Assuming that time

J. Singh, *Sustainability and the Sustainable Development Goals in Cultural Clusters*, Sustainable Development Goals Series, https://doi.org/10.1007/978-3-031-70618-9_9

spent inside C-suites automatically prepares one for a career in policy and politics is reckless. Leadership isn't necessarily replicable from one corporate venue to another and certainly not from one field to the next." This suggests that those who function in the world of business will see issues like sustainability differently from political leaders. The views on sustainability of political and business leaders interviewed for this study are shown in Table 9.1 with similarities highlighted.

National culture can also influence leaders' views on a range of issues including sustainability. Hofstede (2020) extensively studied the dimensions of culture of several countries and found significant similarities and differences among them. In his analysis of Canada, India, Sri Lanka, and the USA he found striking similarities between Canada and the USA (North American cluster) and between India and Sri Lanka (Asian cluster). His database does not include Guyana, the fifth country included in this study. The USA and Canada would be regarded as highly individualist societies while India and Sri Lanka would be collectivist societies. This suggests the likelihood of a predisposition towards sustainability in the latter two countries. On Hofstede's Power Distance dimension, India and Sri Lanka are high while Canada and the USA are low. This may influence the acceptance of sustainability measures if initiated by the state or by business leaders in the Asian cluster and less so in the North American cluster. On Hofstede's Masculinity dimension, Canada, the USA, and India are seen to be masculine while Sri Lanka is seen to be feminine. India being "masculine" is the one instance where it is different from Sri Lanka and like the North American cluster on Hofstede's dimensions of culture. Masculinity would suggest less of a predisposition to sustainability than femininity. The Asian and North American clusters are both moderate to low on Uncertainty Avoidance suggesting some tolerance for risk and in the context of sustainability a disposition to employ innovative policies. On Long-Term Orientation the Asian cluster scores higher than the North American

Table 9.1 Comparison of inter-role themes

	Political leaders	Business leaders
Economic	Long-term profitability	Long-term profitability
	Closing the wage gap	Closing the wage gap
	Sustainability is a responsibility	Growth from support of locals
	Educating the people	Long-term stability
	Adapting to changes	Sustainability as a responsibility
	Making and adapting to change	Adapting to change
		Effort of private sector
		Impact of globalization
		Educating the people
Social	Politics and local culture	Employment equity and fairness
	Income inequity is an issue	Free education provides opportunities
	Employment and local culture	Employment equity and inclusion
	Employment equity	Caste discrimination
	Culture and religion	Employment of women
	Inequity in employment of women	Employment of women in a patriarchal society
	Impact of globalization	Lack of attention on discrimination of ethnic groups
	Government action	Employment of women
	Opportunity gap	Income inequity among races
	Employment of women	Wage and opportunity gap
Environmental	Government action despite costs	Private sector is taking action
	Government taking action	Culture of locals as lacking environmental consciousness

(continued)

Table 9.1 (continued)

	Political leaders	Business leaders
	Everyone's responsibility	The government is leading movement towards environmental consciousness
	Government policies	Effort of private sector
	Influences of Buddhist philosophy	Government policies
	Government action	Changes brought by technology
	Experiences in a developing country	Education of the people
	All countries taking part	Experiences in a developing country
	Consequences of actions	Impact of businesses on resources
		Locals lacking environmental consciousness
		Experiences as a developed country

cluster suggesting a stronger cultural predisposition to sustainability measures. The final Hofstede dimension of culture, Indulgence, is high for the North American cluster and considerably lower for India (data not available for Sri Lanka). High "indulgence" suggests a negative correlation with a culture of sustainability.

Table 9.1 compares themes identified by business and political leaders.

9.1 Overview of Themes from Political Leaders by Country

The sustainability themes identified by political leaders in the countries studied are shown in Table 9.2. As shown in this table, political leaders' views on sustainability could be grouped under the three pillars, which are economic, social, and environmental. Economic sustainability generally involved adapting to changes. In particular, Canadian political leaders claimed that the government was responsible for initiating the changes. The economy was perceived to be drastically changing with the introduction of technology in recent years. Some political leaders believed that one's assets were not limited to physical resources, but also included data. Adapting to such changes was perceived to enhance the economic growth of the country.

In addition, political leaders mentioned employment and income inequity impacting social sustainability. Canadian and American political leaders tended to perceive that women were not experiencing employment inequity as much as they used to, with some women even taking on leadership roles. Political leaders from Guyana, India, and Sri Lanka believed that patriarchal societies were still preventing women from being employed. Some participants believed that expectations of gender roles impacted the employment of women.

The political leaders generally believed that environmental sustainability was influenced by government action. Whether through education subsidies or other financial subsidies, or creating legislation, the government was perceived to have the power to act to foster environmental conservation.

9.2 Overview of Themes from Business Leaders by Country

The themes for business leaders by country are shown in Table 9.3. Business leaders generally agreed with the classification of the three pillars of sustainability, i.e., economic, social, and environmental. As business leaders, the participants perceived that economic sustainability generally involved long-term profitability. Several participants claimed that the sustainability of a business was manifested in its stability. In addition, they believed that economic sustainability also

Table 9.2 Inter-role analysis—political leaders

	Guyana	Canada	India	Sri Lanka	USA
Economic	Long-term profitability	Closing the wage gap	Sustainability is a responsibility	Long-term profitability	Educating the people
	Adapting to change	Making and adapting to changes			Adapting to change
Social	Politics and local culture	Employment and local culture	Employment equity	Inequity in employment of women	Opportunity gap
	Income inequity is an issue		Culture and religion	Impact of globalization	Employment of women
				Government action	
Environmental	Government action despite costs	Government taking action	Everyone's responsibility	Influences of Buddhist philosophy	All countries taking part
			Government policies	Government action	Consequences of actions
				Experiences in a developing country	

Table 9.3 Inter-role analysis—business leaders

	Guyana	Canada	India	Sri Lanka	USA
Economic	Long-term profitability	Closing the wage gap	Growth from support of locals	Long-term stability	Long-term profitability
	Sustainability as a responsibility	Adapting to change	Effort of private sector	Impact of globalization	Educating the people
Social	Employment equity and fairness	Employment equity and inclusion	Caste discrimination	Employment of women in a patriarchal society	Employment of women
	Free education provides opportunities		Employment of women	Lack of attention on discrimination of ethnic groups	Income inequity among races
			Wage and opportunity gap		Lack of political leadership
Environmental	Private sector are taking action	Effort of private sector	Changes brought by technology	Impact of businesses on resources	Experiences as a developed country
	Culture of locals as lacking environmental consciousness	Government policies	Education of the people	Locals lacking environmental consciousness	
	The government is leading movement towards environmental consciousness		Experiences in a developing country		

involves adapting to changes. The majority of business leaders mentioned the contribution of technology and globalization to the economy. Being able to adapt to such changes was linked to the longevity of the business.

Business leaders also believed that sustainability involved social development. The majority of business leaders disregarded gender and race in employment and emphasized the significant contribution of performance. The business leaders generally perceived human resources as assets, and that being able to select from a wider pool of candidates could only strengthen the business.

In terms of environmental sustainability, most business leaders believed that the private sector could impact the protection of natural resources. Company policies involving the conservation of energy or teaching employees to switch to alternative energy sources were small steps corporations could take to impact the environment. Furthermore, some business leaders perceived that choosing sustainable materials and methods reduced the expenses of the company in addition to protecting the environment.

9.3 Similarities and Differences in Leaders' Views by Role and Country

The perceptions of political and business leaders from the five countries studied were compared to each other to identify the similarities and differences in the leaders' beliefs on sustainability. These are summarized in Table 9.4. The majority of the leaders perceived that economic sustainability was driven by adapting to changes and long-term profitability. Social sustainability was associated with employment and income equity. The majority of the leaders discussed the influence of gender and racial minorities in the workforce.

In addition, development status was believed to be associated with sustainability. Most leaders from developing countries—Guyana, India, and Sri Lanka—claimed that globalization had an impact on social sustainability and environmental sustainability. Participants also expressed the view that lower wages in developing countries than in developed countries could be viewed as income inequity. Furthermore, developed countries were believed to have the resources to develop alternative energy, whereas developing countries allocated their resources towards development. Some cultures also influenced sustainability more than others—such as Buddhist philosophy on being one with nature contributing to environmental sustainability among

Sri Lankan leaders, and the caste system resulting in discrimination and employment inequity in India.

In terms of roles, the majority of political and business leaders believed that government action contributed highly to sustainability, especially environmental sustainability. Most of the participants believed that the national governments had the power to create and implement legislation to conserve natural resources and prevent destruction. Some leaders also believed that the private sector influenced sustainability. The majority of leaders were of the view that sustainability was a shared responsibility. Education was seen as a key responsibility that might help people find jobs, gain skills, and utilize resources responsibly.

Table 9.4 Sustainability themes by country and leadership role

	Guyana		Canada		India		Sri Lanka		USA	
	Business leaders	Political leaders	Business leaders	Political leaders	Business leaders	Political leaders	Business leaders	Political leaders	Business leaders	Political leaders
Economic	Long-term profitability	Long-term profitability	Closing the wage gap	Closing the wage gap	Growth from support of locals	Sustainability is a responsibility	Long-term stability	Long-term profitability	Long-term profitability	Educating the people
	Sustainability as a responsibility	Adapting to change	Adapting to change	Making and adapting to change	Effort of private sector		Impact of globalization		Educating the people	Adapting to change
Social	Employment equity and fairness	Politics and local culture	Employment equity and inclusion	Employment and local culture	Caste discrimination	Employment equity	Employment of women in a patriarchal society	Inequity in employment of women	Employment of women	Opportunity gap
	Free education provides opportunities	Income inequity is an issue			Employment of women	Culture and religion	Lack of attention on discrimination of ethnic groups	Impact of globalization	Income inequity among races	Employment of women
					Wage and opportunity gap			Government action	Lack of political leadership	
Environmental	Private sector taking action	Government action despite costs	Effort of private sector	Government taking action	Changes brought by technology	Everyone's responsibility	Impact of businesses on resources	Influences of Buddhist philosophy	Experiences as a developed country	All countries taking part
	Culture of locals as lacking environmental consciousness		Government policies		Education of the people		Locals lacking environmental consciousness	Government action		Consequences of actions
	The government is leading movement towards environmental consciousness				Experiences in a developing country	Government policies		Experiences in a developing country		

References

Bayley J, Kayes C (2018) Why business leaders often fail at politics and policy. Psychology Today, October 8, 2018. Accessed 5 Aug 2020, from https://www.psychologytoday.com/ca/blog/the-helm/201810/why-business-leaders-often-fail-politics-and-policy

Hofstede G (2020) The 6-D model of national culture. Accessed from https://geerthofstede.com/culture-geert-hofstede-gert-jan-hofstede/6d-model-of-national-culture/

Singh J (1984) Education and development: operative and espoused views of the school's role in Guyana. J Can Int Edu 13(2):9–21

Religion: A Fourth Pillar of Sustainability

10

Abstract

Various writers have suggested that the model of three pillars of sustainability—people, planet, profit—is incomplete. They suggest that there may be variables that are not captured in this framework. One such variable is religion. Although it could be argued that it is subsumed under the social pillar there is a strong argument to consider it as a fourth pillar. This chapter explores evidence from the interviews in support of this proposition. It was found that there is a strong link between religious beliefs and sustainability, especially environmental sustainability.

A striking observation in the interviews was the relationship between religion and sustainability. This relationship has been explored by others. For example, Burford et al. (2013) examine the relationship within the proposed pillar of ethical values encompassing cultural-aesthetic, religious-spiritual, and political-institutional variables. Eom and Ng (2023) examined the possible influence of religion on environmental sustainability attitudes and behavior. They categorized religious beliefs on the environment as being dominion or stewardship. "Dominion belief refers to the idea that the Earth was created by God to serve humans, and thus humans have the right to use it for their own needs" (p. 488). This view of dominance of the environment by humans could give rise to the exploitation of the Earth's resources to satisfy perceived human needs even if it results in environmental harm. The opposite is true for the stewardship belief. Eom and Ng (2023) suggest that "the notion of stewardship is commonly found in many, especially, Abrahamic, religions. In these religions, an omnipotent deity is believed to have created the Earth and all creatures in it. He is sovereign over the world. Notably, human beings have a special status, being made in God's image, and entrusted with the responsibility of stewardship" (p. 482). Such beliefs could positively influence attitudes and behavior towards the environment. Eom and Ng (2023) summarize the process linking stewardship beliefs and the environment in Fig. 10.1.

Different groups may cite the Bible to support their stewardship or dominion beliefs. However, the link of such beliefs to environmental behavior follows the process depicted in Fig. 10.1, with perceptions influencing emotions and ultimately behavior. Perhaps to clarify the view of the Catholic Church on the link between Christianity and sustainability Pope Francis in 2015 issued an encyclical entitled Laudato Si. In this document, the Pope "suggests an approach of integral ecology, in which ecological issues are deeply intertwined with economic, social and cultural factors" (Puglisi and Buitendag 2020, p. 7). In the encycli-

J. Singh, *Sustainability and the Sustainable Development Goals in Cultural Clusters*, Sustainable Development Goals Series, https://doi.org/10.1007/978-3-031-70618-9_10

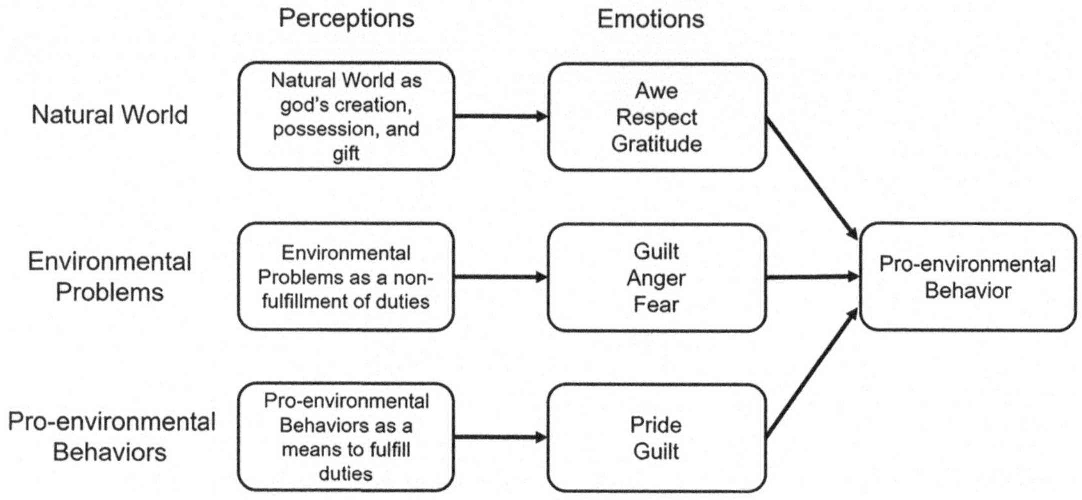

Fig. 10.1 Stewardship beliefs and the environment (Eom and Ng 2023, p. 483)

cal the Pope draws a link among the pillars of sustainability: the environment, society, and economics: "true ecological approach always becomes a social approach; it must integrate questions of justice in debates on the environment, so as to hear both the cry of the earth and the cry of the poor" (Francis Pope 2015, Para. 49).

The stewardship approach to the environment is also held in Islam. Puglisi and Buitendag (2020) illustrate how the environment is linked to Islamic beliefs:

> The Islamic idea of nature is rooted in the Qur'an. It is through the Qur'an that the Muslim faithful perceive their place within the ecosystem. In Islam, God is primarily the God of creation and the religious experience is therefore profoundly marked by a strong relationship between humans and nature. Ramadan, which memorialises the descent of the Holy Qur'an upon the prophet Mohammed, is based on the lunar calendar. We find in the Qur'an more than 750 verses that deal with nature, and 16 of the 114 suras are named after animals or natural phenomena, such as 'the Cow' (2), 'the Cattle' (6), 'the Thunder' (13), 'the Bee' (16), 'the Ant' (27), 'the Daybreak' (89), 'the Sun' (91), 'the Night' (92), 'the Fig' (95), 'the Elephant' (105). (P. 5)

Most of the leaders interviewed for this study in India and Sri Lanka are Hindus and Buddhists respectively. For them, human beings are in a symbiotic relationship with nature and therefore sustainability is inherent in their religious beliefs as they see human beings and nature as being indivisible (Sharma and Tewari 2017).

In this study, the interviewer asked respondents specifically whether their religion influenced their views on sustainability. In the Western nations of Canada, Guyana, and the United States of America, the response was negative, but in the Eastern nations of India and Sri Lanka, the response was almost universally positive. For example, respondents in Canada, while acknowledging a possible link between the two, saw it as an indirect one. Interviewee CB3 explained the link as follows:

> *I don't recall, especially as I was younger and a practising Roman Catholic growing up and going to Catholic schools. I don't remember discussions on the environment—certainly discussions on society, but not necessarily on sustainability, from a sustainability perspective. Well, I definitely think religion has an influence on how we live our lives on a day-to-day basis, as do the focus on wealth or capitalism or success and those things—whatever is important to people. So I do think it has an influence. I guess for me it didn't come to my mind when I thought of sustainability, when I was thinking about sustainability.*

Similarly, Participant CB5 did not see a direct link between religion and sustainability. She explained her view as follows:

I think of those as the social-cultural pieces because my personal view of religion is you are what you are largely because of where you came from in the world and what your parents and family were, not because any one religion is right or wrong.

The indirect link between the two was also echoed by Participant CB2, who was adamant about not being religious but nevertheless grudgingly acknowledged a possible link between religious beliefs and sustainability. He narrated as follows:

Maybe that's a way they get the message, but you know ... I don't know that I'm qualified in that. I think it's a shame that we worship all these gods, and if you look at each religion, it's actually pretty good. I'm not going to say all because I'm not a student of theology, but basically they've got a bunch of good values, and yet we have... religion, just in one community, you've got this fighting this, and like, whoa, man. They're splitting hairs. It can get in the way. A lot of things, battles.

Respondent UP3, a very senior elected State representative in the United States, saw an interesting link between the Christian evangelical movement and their opposition to sustainability. He elaborated on this as follows, describing an example of the dominion view of the relationship between religion and the environment:

We *have another one that I want to be sure that we cover, and that's our problem with our religions.*
Interviewer: *Okay. Tell me something about that.*
UP3: *If you watch the evangelists on television in America, they're primarily creating cults. It's primarily Christian. And it's primarily that you give yourself to Christ and Christ makes all the decisions for you, and Christ will take care of you, Christ will forgive you if you do something wrong, but also we will believe the Bible absolutely unquestioning. And part of the Bible is the end of days. And so, I don't know if you remember a guy named James Watt, who was the secretary of the interior for Ronald Reagan. He's the first one that made me sensitive to this where there's a major portion of the public that thinks all of the natural resources are here for us to use up. To use them up. Not replace them, not protect them, not make the life sustainable. Those are the same people right now that want to make sure that Israel is protected because that's where Armageddon is and it's coming, and they want to enhance it. They want it to come faster, the end of days, so that they can go to whatever their heavenly reward is.*

Participant UP1, an elected municipal leader and executive of the national association of municipal leaders sees the world a bit differently:

Yes, if we believe in the Divine, that we are created in the image of God, and that we have been given all of these resources. The piece that's missing is that we have also been given the brain capacity and the cognitive thinking and the conscience to say, "How do we use the resources in a way that it is not only good for the generation present, but for those in the future?" To say we, the present generation, use up all of the resources and when it is done, then the second coming is here, I don't believe that ... The God I worship is a much bigger God than that, and we have lived centuries, and things have changed immensely. If one travels to other countries and you see the kind of engineering capacity that took place to build all of the beautiful buildings in Rome, in Europe, some of those in India, there's engineering feats that took place hundreds and hundreds of years ago. And now we need huge equipment to do what the human spirit did.

Participant UB4, a business executive who immigrated to the United States from India and is a Hindu saw a link between his religion and sustainability. He explained the link as follows:

What I believe is in our scripture, right from the readers, even things like when we say that, or when we are talking of the Earth as the mother, and worshiping; it's not necessarily worshiping in the sense that you create a deity and you do it. It's the actual worship outside of the ritualistic aspect. It's to take care of it, which I feel is connected back to sustainability. What I found is the basic tenets of these religions are very similar. Obviously, the rituals apart, so it's essentially an actualization of Hinduism in the various forms based on those ages more than anything else. I'm not able to figure out a particular verse, which we can talk about it, but when I think Mother Earth, it's not about just doing a lip service to it. It's about that sustainable aspect, which is woven into the Indian culture, I would say. Take just the cow. That's a great example of sustainability once you think about it. Right from the mouth, to even the cow dung. That is supposed to be one of the central pillars of our breathing society of Hinduism. Everything kind of revolves around the animal, but the fact there is that everything is just so usable there. Again, I think it is woven into the culture. Again, Hinduism, depending upon it's a way of life. It's not necessarily an organized religion from the perspective, but I feel that it is there.

The views of Participant UB4, a Hindu and Indian immigrant to the United States, provide a look at the unquestioning link between religion and sustainability. His is a stewardship view of that link that was also found among respondents in India and Sri Lanka. Participant IP6, a senior Indian public administrator addressed the link between religion and economic sustainability. He spoke of the teaching of Mahatma Gandhi on the subject:

> *We will become heavily geared towards economic things. That is happening, unfortunately. I think this is against the Indian philosophy; it's against the Indian way of life. In India, we always believe in what Gandhi told, that you take as much as you need, not beyond that. That is philosophies, that we talk of Sikhs, we talk of Hindus, we talk of Buddhists, but you'll find one thing very common in all these. Everywhere they've told don't accumulate more than what you need. Because, if you have more than you need that turns into greed, and greed leads to more kind of problems—political, social, economic—in this society. So what I would say is that this is causing unrest in the society.*

Participant IP4, a minister in an Indian State government directly addressed the link between Hinduism and public policy, specifically, environmental pollution*:*

> *In the Hindu religion, we believe that we should protect the environment. We teach our children this, and that's part of how we live. We believe that trees have life, everything has life, so you can't destroy them. Because if you destroy the natural environment, you also will not survive. When we are preparing food, for example, we prepare for us, for the cow, and then the last is for us. These beliefs drive government policy.* Interviewer: *These beliefs drive government policy? These beliefs drive government policy, yes, yes. These play a part in policy of the central government and of the state government. We think that's on the basis of these philosophies we can develop a sustainable society.*

Participant IP4 was clear in terms of describing the influence of religious beliefs—not only on personal views on sustainability but also on government policy at both the state and national levels. Participant IP3, a very senior elected official in a large Indian state, who previously also served at the national level, narrated his ideas on the link between Hinduism and sustainability*:*

> *As far as sustainability is concerned, this process continues, and changes are also there. The present environment in which we are living now, there are good people in this country and also follow good standards, practices, so let's say that there are people with a strong determination of their thoughts.*
>
> *But both good and bad people find everywhere. Therefore, if anything has formed, then it's more necessary to keeping it continuous, but it also have subject to changes.*
>
> *Like there are many religious beliefs, measurement standards, and many rules are associated to religion and many types of things get changed from time to time.*
>
> *Now, like in our Hindu religion, we offer water to the Sun as belief of 'Lord Sun', then it became religious practices. But it's not only religious belief; we applied science as well, because when we offer water to the sun, the radiation of sunlight passes through water and then the light gets reflected in our eyes, and it enhances the eyesight. This is the scientific way but included in religion. In the same way before starting cooking of food, our priest sprinkles water on the ground. Its religious belief is that it has become pure, whereas there is scientific view for that: earth surround with carbon dioxide and carbon dioxide dissolves in water, so when we sprinkled water, then the carbon dioxide gets dissolved in water and only oxygen was left over. And this makes food purified, and this practice has been followed continuously. There were chances in the beginning that people would believe or would not believe if it is being said by someone that it would be purity by sprinkling water. In order to bring continuity, people make it as form of religious practice, which people accept it and follow too.*
>
> *For continuity, it is not necessary that people always accept things by law and rules—regulation. Sometimes it is because of somebody's fear or some greed mould them. So if we change them by their thinking and making a religious belief, then they will do. And it will be sustainable also. So my point is that it has been happening from time to time.*
>
> *When it comes to environment, then we make the environment deteriorate ourselves. Earlier the earth was surrounded by trees. For our comfort and selfishness we cut down trees and which reduce the plants on the earth, due to which the rain began to subside, leads to weather changes. It also changes the climatic cycle, now as compared to earlier rain water level reduce. Winter ends by February; earlier it was till March. Whenever a change happens in nature, the human being has a big role in it. He changes the nature, therefore the change comes from the nature.*
>
> *We want to communicate that there let's be a sustainable development, but do not harm the nature.*

Development is a continuous process and keeps moving. According to Hindu mythology, nature destroyed everything; only "Manu and his wife" lived and the rest of the earth would have gone. Now again, whole universe has grown and with huge population. Development would allow but it should follow the principle of sustainability, otherwise nature will do balancing by its own. My point is to say that, population is increased with great number, so it is necessary to take some action to control and make balance in it. For that, government of India has run a campaign of sterilization, but it does not make any difference to it, but nature played a huge role in it—like tsunami came, 200,000 people die, two Island of Andaman Nicobar have ended.

So I say nature is changing, and nature has a power to change everything in environment. Man can't control every change made by the environment, but yes, in order to save the nature, we can plant trees to control degradation.

Similarly in Sri Lanka, most of the respondents saw a link between Buddhism and sustainability. Participant SP3, a member of the Sri Lankan parliament and a former government minister, was forthright in linking Buddhism to sustainability and public policy. He elaborated on the relationship as follows:

Well, actually sustainability has been a novel concept for Western world. It's not something new to us. Being Buddhist, so Sri Lanka has been a country with Buddhist civilization, and Sri Lankans are considered. Buddhism has been learned as a philosophy in a certain part of the world. And in other parts of the world, they consider Buddhist rituals. But here we have transformed the Buddhist philosophy to fit a lifestyle.

So most of the religions say that the world has been created by the God for the man. So man in it, and he started to utilize the resource created for him; he destroyed the entire environment. But in Buddhism says, "Man is only a part of the environment, which is interconnected, interwoven, and interdependent." So change in environment has a drastic reverse impact on the man.

So these concepts have been discussed in this world, and they always got theatre. But hundreds of years before theatre, Buddha talked about the environment. There's a classic dialogue between Arahat Mahinda, who brought Buddhism from India to Sri Lanka around 2,300 years ago. In fact, Arahat Mahinda is the well-known Emperor Ashoka's one and only son. So, in a dialogue with a reigning Sri Lankan King, Arahat Mahinda says, "Oh King, you are not owner of this land. We are not owner of these trees, animal, or birds who are flying all over. You are merely the custodian who poses on behalf of others. So you have to look after; you should ensure these resources will be used for everybody's benefit, including the trees and the birds. And you should further ensure forthcoming, you will utilize the resources in such a way that forthcoming your impression will also be able to use in the very same manner that you use." Isn't it all about sustainability?

Other Sri Lankan respondents echoed Participant SP3's views. They overwhelmingly articulated the belief that damaging the natural environment is contrary to their religious beliefs.

References

Burford G, Hoover E, Velasco I, Janouskova S, Jimenez A, Piggot G, Harder MK (2013) Bringing the 'missing pillar' into sustainable development goals: towards intersubjective values-based indicators. Sustainability 5(7):3035–3059

Eom K, Ng ST (2023) The potential of religion for promoting sustainability: the role of stewardship. Top Cogn Sci 15:480–499

Francis Pope (2015) Encyclical letter LAUDATO SI' of the Holy Father Francis on care for our common home, viewed 09 April 2020, from https://w2.vatican.va/content/dam/francesco/pdf/encyclicals/documents/papa-francesco_20150524_enciclicalaudato-si_en.pdf

Puglisi A, Buitendag J (2020) The religious vision of nature in the light of Laudato Si': an interreligious reading between Islam and Christianity. HTS Teologiese Studies/Theol Stud 76(1):a6063. https://doi.org/10.4102/hts.v76i1

Sharma E, Tewari R (2017) Teaching sustainability through traditional wisdom: a conceptual framework for business studies. Curr Sci 113(2):236–244. http://www.jstor.org/stable/26293855

11 Conclusion

Abstract

As 2030 approaches it is becoming clear that the UN sustainability goals will not be achieved in their entirety. Perhaps just a handful of objectives will be achieved. Does this mean that the world community will have to go back to the drawing board to address the sustainability imperative? Perhaps not, but political and business leaders will have to refine their approach to the challenge. This chapter considers some of the issues to be addressed in moving forward.

Sustainability and sustainable development entered discussions of world development with the Brundtland Report in 1987. Subsequently, John Elkington in 1994 addressed the issue as it relates to business with his three pillars, triple bottom line, approach: People, Planet, and Profit. Then in 2015, the United Nations General Assembly adopted the 17 sustainability goals (SDGs) and set in motion an ambitious program for the achievement of the goals by 2030.

This book examined the above-mentioned and related concepts in a qualitative study of the views of business and political leaders on sustainability in five countries: Guyana, Canada, India, Sri Lanka, and the USA. These policymakers and influencers are critical to the achievement of the SDGs in their countries. However, only a handful of the 41 leaders interviewed for this study were even aware of the SDGs, and fewer knew of the Brundtland Report or John Elkington. However, as demonstrated in the analysis of the interviews, they have thoughtful views on sustainability and various themes run through their opinions. Analysis of the interviews with the 41 business and political leaders yielded the themes on sustainability that were discussed in the foregoing chapters. Moreover, the interviews further provided insight into several issues related to the concept. Some of these issues are examined below.

11.1 Re-examining the Definition of Sustainability

The Brundtland Report (1987) provides what has become the core of most definitions of sustainability. The report defines sustainable development as "development that meets the needs of the present without compromising the ability of future generations to meet their own needs" (p. 41). Implicit in this definition is that there are fundamental needs for people to exist. According to the report, for the poor, these are essential needs and must be of foremost concern. Also implicit in the definition is that the world's resources are not limitless, and therefore, there

J. Singh, *Sustainability and the Sustainable Development Goals in Cultural Clusters*, Sustainable Development Goals Series, https://doi.org/10.1007/978-3-031-70618-9_11

must be control of what is done in the present so that resources will be available for future generations.

The definition also raises the notion of intergenerational equity as resources must be currently used in a way that future generations will be able to meet their needs. The report suggests that this concern for intergenerational equity implies intragenerational equity. However, not everyone believes in intergenerational equity. While most of the participants in this study acknowledged a responsibility to future generations, some were dismissive of the principle, such as participant SB2 (a prominent Sri Lankan entrepreneur) in an exchange with the author.

> Interviewer: *If we go back to the meaning of sustainability—that sustainability means using our resources in such a way that future generations will have access to these resources. Could you give me an overall comment on that?*
> SB2: *I cannot answer that question because my thinking is entirely different from your perception.*
> Interviewer: *But tell me your thinking.*
> SB2: *My thinking is this all depends on the global situation. That particular country situation, that particular part you are dealing with, so you cannot directly say this will be preserved for your future generation, because I myself don't know that this can continue or may not be continued. So all these things are washed away. So the things that are happening, how can you give a guarantee that this can be sustained for the future growth of the people and the generation. We have to now live for the day and live for the week. The things are every day changing and so fast.*
> Interviewer: *Tell me more about living for the day, for a week, for a year.*
> SB2: *We go on to take decisions for the day. You cannot take decisions based on certainty... look at the agricultural product. We know this month you will have rain, we know this month you will have olive, next month you will have production of tea. If it doesn't rain, then there is no production of tea. See what happened to Sri Lanka now. Eight months no rain. It is practical. It is not theoretical.*

While a few participants scoffed at the notion of a responsibility to future generations, many did not even locate the idea in their conception of sustainability. For many, sustainability was seen as being able to continue current activities in the political and business spheres. Participant GP4, a Guyanese political leader, in his narrative on the meaning of sustainability, illustrates a common understanding of sustainability to mean longevity of economic or political activities. He explained as follows:

> *Well, sustainability in economics, I'm not too familiar with it. Sustainability became an issue in economic development sometime in the nineties or in the eighties. Guyana became affected by [it], in the sense that aid agencies required all programs that had lengthier assistance to be sustainable. That was a primary requirement. I do not understand the technical or the academic meaning of sustainability, but they wanted programs which they funded to have a degree of sustainability so that when they left or when the program was completed, the impact of it would remain for the long term. We received a great deal of assistance from aid agencies. This was a fundamental principle which applied to all of our programs, so that they don't give you money to buy a desk, but they give you money to send somebody for training. That is how I understood it.*

A similar narrative emanated from Participant UB2, an American business leader. In his words:

> *Sustainability to me would mean, how do we sustain our organization to be able to do what's our vision and our mission? In health care, clearly, we are here to provide health care for our community, improve the health of our community, and to do it in a manner that sustains our organization. So it's also good for our employees, good for our physicians, good for the organization. And how do we do that in a manner that can be sustainable really forever? And I think a little bit different from maybe some businesses and our business of health care—maybe it's just the hat I wear. I think sustainability of health care. We know health care is not going away. So we need to be able to sustain it and improve on it in hope, but in reality it's ... we're going to sustain it. Are we going to sustain it in a manner that makes it improved and better or, in some cases, sustainability in health care says how do we adjust and maybe do it differently?*

For both Participants GP4 and UB2, a developing country political leader and developed country business leader respectively, sustainability has nothing to do with the use of resources in such a way that the needs of future generations will be met. Their primary concern was the survivability of organizational activities. This is not to suggest that they do not care about the needs of future generations, but their understanding of sustainability is not as defined in the Brundtland Report.

11.2 The Interconnectedness of Sustainability

In the discussion of the concept of sustainable development, the Brundtland Report addressed the issue of equity and fairness concerning those living in developing countries. "The satisfaction of human needs and aspirations is the major objective of development. The essential needs of vast numbers of people in developing countries for food, clothing, shelter, jobs—are not being met, and beyond their basic needs these people have legitimate aspirations for an improved quality of life. A world in which poverty and inequity are endemic will always be prone to ecological and other crises. Sustainable development requires meeting the basic needs of all and extending to all the opportunity to satisfy their aspirations for a better life" (Brundtland Report 1987, p. 42).

The report emphasized the interconnectedness of societies everywhere, highlighting the fact that excessive consumption in one part of the world can be detrimental to the sustainability of the entire planet. This implies that growth and development everywhere must be considerate of the needs of all. Actions that do not respect this principle would therefore threaten sustainability. "In essence, sustainable development is a process of change in which the exploitation of resources, the direction of investments, the orientation of technological development, and institutional change are all in harmony and enhance both current and future potential to meet human needs and aspirations" (Brundtland Report 1987, p. 43). Some of the political leaders in the developing countries were cognizant of the need for equity and fairness between the developed and developing world. Participant GP1, a high-ranking political leader in Guyana, offered his view of the relationship:

> *I think it's very important because right now we see the World Trade Organization being questioned. I remember, I don't know if you were, but I was doing my O-level exam in the times when we were still a colony, and one of the common questions used to come in history those days was: "Did trade follow the flag or did the flag follow trade?" And you're supposed to discuss: look at the British Empire, the growth of it, and so on. Did trade follow the flag? If I had to answer that question now, I would say it doesn't matter which one follows which. The important thing is that they are always close together. You can't have trade without ideas of common citizenship and so on starting to get into it. I think the WTO would also be; I think it would continue to progress, but it would progress limited by the people realizing and experiencing a common humanity kind of thing. I think the two things, they can't be too far apart. They can't be too far apart.*

Participant GP3, although from a different political party than GP1, relied on a shared experience to express similar sentiments:

> *So the concept of sustainable development cannot be a slogan. Because everyone is talking now about sustainable development. But the question is that you have to look at sustainable development in the context of a global approach. You cannot have a situation where 5% of the world dominate/ consume 85% of the resources of the world. That's unsustainable. You cannot have a world where 5% of the world pollutes 80% of the atmosphere and damages the ozone layer. You cannot have a world where 5% of the world decides they gonna have climate change fuelled by the decimation of forests, or by the melting of our northern and southern ice. You can't have that situation of dis-proportionality, where some feel that they can have access to resources that they can use any way they feel. While the effect of that is that other countries are faced with problems like flooding, Guyana is 5 feet below sea. If you have a rise in the sea level, we would be submerged.*

Knowing what needed to be done to create a semblance of equity between the developed and developing countries, the Brundtland Commission identified strategies to achieve its objectives. "In its broadest sense, the strategy for sustainable development aims to promote harmony among human beings and between humanity and nature" (Brundtland Report 1987, p. 57). The objective here is twofold: the fostering of equity between the developed countries and the developing countries, while at the same time promoting responsible environmental practices. To this end, the Brundtland Commission identified seven initiatives. The first of these was the revival of growth. This was in reaction to a period of slow economic growth in the 1980s. However, this may again become an issue as COVID-19 has wreaked havoc on the world's economy and would exacerbate economic inequities within and among nations.

The second initiative identified by the commission qualified the first objective on economic growth by advocating for changing the quality of growth. As stated in its report, "These changes are required in all countries as part of a package of measures to maintain the stock of ecological capital, to improve the distribution of income, and to reduce the degree of vulnerability to economic crises" (Brundtland Report 1987, p. 48).

The third initiative identified in the report is that of meeting essential human needs. The argument here is that even if goods and services are available in a country, poverty will prevent some from meeting their essential needs. Key to meeting essential needs is the availability of employment opportunities.

The fourth initiative is a recognition that there is a limit to population growth and measures would have to be taken to ensure a sustainable population level. Of course, this will always be a controversial area, but the Commission noted that it is not only the growth in the world's population that is important but also the geographic growth patterns. It has commonly been argued that the standard of living in developed nations places a heavier burden on the environment than comparable populations in the developing world.

The fifth initiative in the Commission's strategy for sustainable development is the conservation and enhancement of the resource base. This is central to the philosophical underpinning of the report's argument for intergenerational equity as use of the planet's resources must be subject to the needs of future generations.

The sixth initiative identified in the Report is directly related to the conservation of resources and addresses the issue of reorienting technology to the needs of the developing world and the environment. "First, the capacity for technological innovation needs to be greatly enhanced in developing countries so that they can respond more effectively to the challenges of sustainable development. Second, the orientation of technology development must be changed to pay greater attention to environmental factors" (Brundtland Report 1987, p. 54).

The final initiative advocated for a merging of environmental and financial considerations in decision-making. Thus, the ecological impact of decisions made by governments and business must be considered alongside the economic ramifications.

The simultaneous consideration of economic as well as other factors in sustainable development was echoed in the highly influential depiction of sustainability advanced by John Elkington. In his 1997 book, *Cannibals with Forks*, Elkington sought to offer a way to reform capitalism to ensure its survival. His solution was for businesses to focus not only on profits but also on the physical environment and social justice. Hence, he suggested that the performance of a business be evaluated on the pillars of profit, planet, and people. The three bottom lines are seen to be constantly changing, thus presenting a challenge for business as the social, economic, and environmental pillars cannot be viewed in isolation. Many variables formerly viewed as external to the management process now must be factored into the management process and a business' performance evaluation. This approach has attracted considerable attention and generally guided the interviews conducted for this study.

11.3 Measuring Sustainability

Elkington and others have identified the need to be able to measure what one wants to manage. In the case of sustainability, therefore, there would be the need to measure company performance on financial, environmental, and social matters. Metrics for assessing the financial performance of firms have long existed, but measurement of performance on social and environmental indicators has been a challenge for firms. Nevertheless, numerous companies today report on their triple-bottom-line performance.

Soon after the publication of *Cannibals with Forks* in 1997, Elkington wrote on the efforts of Shell to address the three pillars of sustainability.

Citing Shell's report, he drew a parallel between the triple bottom line and Shell's efforts to address additional responsibilities: "Earlier this year, Shell, the world's largest oil company and one of its longest surviving multinationals, issued a report about its values." With the challenging title *Profit and Principles—does there have to be a choice?* the Anglo/Dutch group set out to describe "how we, the people, companies, and businesses that make up the Royal Dutch/Shell Group, are striving to live up to our responsibilities— financial, social, and environmental" (Elkington 1998, p. 18).

How then does the firm measure its performance on the three bottom lines? Elkington notes that accounting for finances has a long history but is still deficient. Social responsibility probably has as long a history as business itself, but accounting for it is not well established. Environmental responsibility has assumed critical importance considering the fragility of our planet and mounting evidence that there is a direct link between the poor state of the environment and business activities. However, the lingering question is: How does the firm account for the environmental impact of its activities?

Elkington notes that the economic bottom line may be addressed through various financial ratios. "A company's bottom line is the profit figure used as the earnings figure in the earnings-per-share statement, part of standard accounting practice. In trying to assess a company's conventional bottom line performance, accountants pull together, record, and analyze a wide range of numerical data" (Elkington 1998, p. 19). He suggests that key questions on economic sustainability would be whether the company's innovation would keep it competitive in the long term, and whether it would be able to retain its valuable human resources. In the interviews with political and business leaders, this conception of sustainability was often mentioned. Participant GB1 captures the sentiment:

> *Immediately what comes to mind is the long-term profitability, taking the long view and tying it in with ethics. It obviously means, I would think, having a very conservative view of your strategies, not looking for the quick buck, as we say. But looking to position your business so that over the long-term, you can continue to add value, build value, for your stakeholders.*

How does a firm measure performance on its environmental bottom line? This pillar of sustainability is not as easily ascertained as economic performance. However, environmental accountability is indirectly obtained by state regulators. "Some countries, like The Netherlands, also back up their regulations with voluntary programs designed to push companies towards sectorally agreed targets. Just as often, however, business is held to account by environmentalist and media campaigns, which may bear little relation to regulated or voluntarily agreed targets" (Elkington 1998, p. 21). There are also voluntary measures that companies could take such as ISO 14000 certification by the International Organization for Standardization.

The social pillar of Elkington's conception of sustainability perhaps has a history as long as commerce itself. The impact of commercial activities on society has always been known, but the question that has not always been answered is whose responsibility it is. Debates on the social responsibility of business have been ongoing for centuries with some arguing that business has no social responsibility and should be focused solely on making profits. Others have argued that as an institution of society, granted a license by the state, it is the primary role of business to serve the needs of society, and when it ceases to do so its license should be revoked. The social responsibility of business has been emphasized in the Brundtland Report, the Sustainable Development Goals, and Elkington's triple bottom line.

Again, as with environmental accounting, the question of measurement becomes an issue. How does a firm measure its social performance? Elkington (1998, p. 22) suggests that: "Social accounting aims to assess the impact of an organization or company on people both inside and outside. Issues often covered are community relations, product safety, training and education initiatives, sponsorship, charitable donations of money and time, and employment of disadvantaged groups." A common dilemma for business leaders is that they recognize social problems and the possible role business plays in creating them and perhaps even in solving them but offer no concrete way in which business can identify and measure its social performance. The comments of Participant CB5 on the issue are instructive:

> *There's a* Macleans *article that talked about, it was an interesting one where they compared the status of Aboriginal people to Black people in the US and in every category, quite frankly, the status of the Aboriginals was worse off: life expectancy, average income, participation in the workforce, things like that. Clearly there's work to be done there. I'll be honest, I have no idea what the answer is; I just know it's an issue. I'm not a social scientist on this thing, and I think it's really complicated, both historically and socially. I think many Canadians are quite self-congratulatory on this topic in saying that we are wonderful, we're perfect, we're so much better than everyone else on this thing. Maybe we are a little bit, but I don't think we're perfect by any stretch of the imagination. I still think there's groups of people who are unfortunately disenfranchised.*

11.4 Beyond the Three Pillars?

Most of the themes that emerged in the interviews could be located within the economic, social, and environmental pillars of sustainability. However, there are themes that warrant separate classification as the traditional three pillars are not exhaustive; important determinants of sustainability are left out of Elkington's three pillars. For example, as discussed in Chap. 10, a finding of this study is the importance of religion in determining attitudes to sustainability and its possible inclusion as a fourth pillar of sustainability. Of course, this would raise even more questions about measurement, but there are measures of religiosity. It could be argued that the impact of religion is already reflected in the social, economic, and environmental pillars, but some would argue that religious beliefs and practices embody sustainability, and therefore someone observant of such a religion would foster sustainability. Participant SP3, a senior Sri Lankan political leader, explained that Buddhist practices define environmental sustainability:

> *And they consider damaging environment as a big sin, because elderly Buddhist monks, senior Buddhist monks are not supposed to take a leaf out of the tree. Because Buddha says, "Trees are living, although they can't move." So monks should not even take a leaf out of it, because the tree would feel about it. And they're not supposed to cut trees anyway. But senior monks, they're not even supposed to take a leaf out of it.*

In a similar vein, Burford et al. (2013) argue for the inclusion of a fourth pillar of sustainability that reflects human values, ethics, and worldviews. They recommended that: "The institutions tasked with developing Sustainable Development Goals, targets, and indicators should take time, especially at this critical juncture, to reframe the sustainability assessment process and incorporate an explicit acknowledgement of its ethical dimensions" (Burford et al. 2013, p. 3053).

11.5 Examining Progress with the Sustainability Development Goals

The Sustainability Development Goals may be viewed as an operationalizing of the triple bottom line as it applies to global development. The 17 goals are classified as economic, social, and environmental, and while it may be argued that they flow from the Brundtland Report, they also reflect the objectives of the triple bottom line. As a conclusion to this book on sustainability, it is therefore apt to briefly revisit the progress being made on the 17 SDGs with less than a decade left for their achievement. Progress towards the achievement of the goals has been affected by the COVID-19 pandemic, and it now appears as if many of the objectives will not be achieved by 2030. Of course, much could happen in the next few years to accelerate progress toward the goals, but in the absence of resolute action by the world community, the laudable measures aimed at creating a sustainable future for mankind appear to be in jeopardy.

As 2030 nears it is becoming clear that the totality of the SDG targets will not be attained. There will be uneven results among nations and on different targets within nations. The impact of this on the five countries where interviews for this book were conducted with business and political leaders would vary as they face different sustainability challenges. In Guyana, originally included as a pilot study but providing rich data on ethnic relations and fossil fuels, the future is uncertain. In national elections held in March 2020, there was much acrimony, and the results are still being contested in the courts. Those elections were basically a battle between the country's two major ethnic groups, Afro-Guyanese and Indo-Guyanese, represented respectively by a Coalition of parties and the People's Progressive Party. This is clearly unsustainable.

What is also unsustainable is a "split personality disorder" in Guyana's development policies. The country maintains that it is pursuing a green economy but is a new petrostate with ExxonMobil committed to pumping billions of barrels of oil from deep below the sea some 160 kilometers off the Atlantic coast of the country. Asked about this in the interviews, Guyanese leaders suggested that they see no environmental issues with operations 160 kilometers away from where people live. What might they say about ExxonMobil flaring tens of billions of cubic feet of natural gas in their Guyana offshore oil production? What would they say about fossil fuels produced in Guyana and used elsewhere contributing to global warming? Would they still deny the impact on their and the planet's well-being?

India, the world's most populous nation, where interviews were also conducted is the world's largest democracy and its third largest economy. Thus, India's future will be pivotal to the planet's future. However, it is a country beset by stubborn sustainability challenges. India is home to some of the most polluted cities on the planet. New Delhi, the seat of India's central government, is the most polluted city in the world. Moreover, the country faces severe problems relating to religious conflict and income inequality. This contrast between the rich and poor is visible in many Indian cities, but nowhere more so than in Mumbai which boasts the world's most expensive private residence, the US$2 billion Ambani home, juxtaposed against sprawling slums that house half the city's population. How will India overcome its sustainability challenges?

In nearby Sri Lanka, the resplendent island, the pollution problems are not as severe as in India, but the country continues to grapple with the sustainability challenge of a bitter ethnic conflict between the Singhalese majority and the Tamils, who in the past have taken up arms to seek a nation of their own. Moreover, how will they manage an economy that is burdened by a heavy debt load fuelled by loans from China for infrastructural work done by Chinese companies?

In the two North American countries where interviews were conducted, the USA and Canada, the grinding poverty seen in Guyana, India, and Sri Lanka is not a sustainability challenge, but wealth and income inequality beset both countries. The USA, often described by its leaders as "the greatest country on earth," failed miserably in its management of COVID-19. Moreover, the Trump administration rolled back several environmental protection measures and took the country out of the Paris climate agreement. The latter was reversed by President Biden. On the social front, the state of race relations in the USA has again come under scrutiny, as riots broke out in several cities in protest against police treatment of minorities. However, a new term, "systemic discrimination," has crept into the vocabulary of policymakers and perhaps will help in overcoming this challenge to sustainability.

In nearby Canada, a country sometimes described as the "Peaceable Kingdom," systemic discrimination has long been in the vocabulary of academics and is now increasingly used by political and business leaders. However, talk is cheap and despite recognition of systemic discrimination against minorities in Canada, policymakers have done little to address the problem.

They have also done little to address the challenge of income and wealth inequality. As the economic impact of COVID-19 brought to light the suffering of working people, the government of Prime Minister Justin Trudeau pledged to address the issue of income inequality and specifically mentioned the concept of a guaranteed minimum income. This announcement was quickly followed by reports of big business becoming concerned about a ballooning budget deficit and growing national debt. Foremost among those voicing concerns were the CEOs of Canada's big banks that have profits of tens of billions of dollars per quarter. The talk of a guaranteed minimum wage is now just a whisper emanating from the New Democratic Party which holds just 24 of the 338 seats in Canada's federal parliament.

What was evident in the interviews in all the countries is that political leaders, contrary to expectations and with some exceptions, lacked the vision and will to create a sustainable planet. The business leaders, with few exceptions, were not hostile to repositioning their strategies to embrace the social, economic, and environmental pillars of sustainability. Perhaps the leadership to ensure a sustainable future has been ceded to business by government.

An example from Canada illustrates this point. In November 2020, "The chief executive officers of eight Canadian pension funds, collectively representing about $1.6 trillion in assets under management, *called* for a green recovery from the COVID-19 economic slump" (Langerquist 2020). The CEOs appealed to investors to make sustainability the focus of the expected economic recovery. The governor of the Bank of Canada, Tiff Macklem, expressed his support for the position of leaders in the financial sector. He stated, "A strong commitment to environmental sustainability, diversity and inclusion and good governance principles will not only make our economy and financial system more resilient, it's also the right thing to do. Leadership from Canada's financial sector is essential as we focus on building an enduring and more equal economic recovery from the pandemic" (Langerquist 2020).

However, even if business, as in the Canadian example cited, embrace sustainability it is virtually impossible that it will be a coordinated effort with the same goals measured by global standards. A sustainable world will only be realized by a coordinated global effort by governments. It is for this reason that the UN's Sustainable Development Goals are critically important.

Unfortunately, it is unlikely that the SDGs will be attained by 2030. Now that COVID-19 is seemingly under control and the world can breathe again, the United Nations will redouble its efforts to ensure that the world's resources are used in a fashion that will ensure that future generations will have access to resources to meet their needs. However, the key to a sustainable future is the education of citizens so they will elect governments committed to sustainability, but that is a conversation for another day.

References

Brundtland Report (1987) Our common future: report of the World Commission for Environment and Development. WCED, New York

Burford G, Hoover E, Velasco I, Janouskova S, Jimenez A, Piggot G, Harder MK (2013) Bringing the 'missing pillar' into sustainable development goals: towards intersubjective values-based indicators. Sustainability 5(7):3035–3059

Elkington J (1997) Cannibals with Forks: the triple bottom line of 21st century business. Capstone Publishing Ltd, Oxford

Elkington J (1998) Accounting for the triple bottom line. Meas Bus Excell 2(3):18–22

Langerquest J (2020) Canadian pension funds representing $1.6T AUM call for green COVID-19 recovery. Accessed 25 Nov 2020 from https://ca.finance.yahoo.com/news/canadian-pension-fund-giants-call-for-green-covid-19-recovery-132358794.html

Appendix

J. Singh, *Sustainability and the Sustainable Development Goals in Cultural Clusters*, Sustainable Development Goals Series, https://doi.org/10.1007/978-3-031-70618-9

Appendix 1: Frequency of Mention of sustainability Pillars by Country and Role

	Guyana		Canada		India		Sri Lanka		USA		
	Political $n = 5$	Business $n = 4$	Political $n = 3$	Business $n = 5$	Political $n = 6$	Business $n = 4$	Political $n = 3$	Business $n = 3$	Political $n = 4$	Business $n = 4$	Total $N = 41$
Economic	14	18	13	18	18	9	5	9	12	15	131
Social	19	17	5	16	6	11	12	19	10	13	128
Environmental	6	20	11	19	6	16	24	4	12	10	128
Total	39	55	29	53	30	36	41	32	34	38	387

References

Bayley J, Kayes C (2018) Why business leaders often fail at politics and policy. Psychology Today, October 8. From https://www.psychologytoday.com/ca/blog/the-helm/201810/why-business-leaders-often-fail-politics-and-policy. Accessed 5 Aug 2020

Bruntland Report (1987) Our common future: report of the World Commission for Environment and Development. WCED, New York

Burford G, Hoover E, Velasco I, Janouskova S, Jimenez A, Piggot G, Harder MK (2013) Bringing the 'missing pillar' into sustainable development goals: towards intersubjective values-based indicators. Sustainability 5(7):3035–3059

Carpar DV, Devinney TM, Kirkman BI, Caligiuri P (2015) Conceptualizing and measuring culture in international business and management: from challenges to potential solutions. J Int Bus Stud 46(9):1011–1027

Chitiga M, Sekyere E, Tsoanamatsie N (2014) Income inequality and limitations of the Gini index: the case of South Africa. Human Sciences Research Council, Pretoria. From http://www.hsrc.ac.za/en/review/hsrc-review-november-2014/limitations-of-gini-index. Accessed 24 July 2019

CIA (2019) The World Factbook: Guyana. From https://www.cia.gov/library/publications/the-world-factbook/geos/gy.html. Accessed 5 Feb 2019

CIA (2020a) The World Factbook: India. From https://www.cia.gov/library/publications/the-world-factbook/geos/in.html. Accessed 25 July 2020

CIA (2020b) The World Factbook: Sri Lanka. From https://www.cia.gov/library/publications/the-world-factbook/geos/in.html. Accessed 29 July 2020

CIA (2020c) The World Factbook: USA. From https://www.cia.gov/library/publications/the-world-factbook/geos/in.html. Accessed 30 July 2020

CIA (2023a) The World Factbook: Canada. From https://www.cia.gov/library/publications/the-world-factbook/geos/ca.html. Accessed 19 July 2023

CIA (2023b) The World Factbook: USA. From https://www.cia.gov/the-world-factbook/countries/united-states/. Accessed 6 Sept 2023

CIA (2023c) The World Factbook: India. From https://www.cia.gov/library/publications/the-world-factbook/geos/in.html. Accessed 25 July 2023

CIA (2023d) The World Factbook: Sri Lanka. From https://www.cia.gov/library/publications/the-world-factbook/geos/in.html. Accessed 20 July 2023

CIA (2023e) The World Factbook: Guyana. From https://www.cia.gov/library/publications/the-world-factbook/geos/gy.html. Accessed 20 July 2023

Davies J, Lluberas R, Shorrocks A (2018) Credit Suisse global wealth databook. Credit Suisse, Zurich

Davies J, Lluberas R, Shorrocks A (2022) Credit Suisse global wealth databook. Credit Suisse, Zurich

Davies J, Lluberas R, Shorrocks A (2023) Credit Suisse global wealth databook. Credit Suisse, Zurich

Denzin NK, Lincoln YS (eds) (2005) The sage handbook of qualitative research, 3rd edn. Sage Publications Ltd., Thousand Oaks

Diaz S, Demissew S, Carabias J, Joly C, Lonsdale M, Ash N et al (2015) The IPBES Conceptual Framework—connecting nature and people. Curr Opin Environ Sustain 14:1–16

Elkington J (1997) Cannibals with forks: the triple bottom line of 21st century business. Capstone Publishing Ltd, Oxford

Elkington J (1998) Accounting for the triple bottom line. Meas Bus Excell 2(3):18–22

Elkington J (2018) 25 years ago I coined the phrase 'triple bottom line'. Here's why it's time to rethink it. HBR, June 25. From https://hbr.org. Accessed 30 Jan 2019

Eom K, Ng ST (2023) The potential of religion for promoting sustainability: the role of stewardship. Top Cogn Sci 15:480–499

Francis Pope (2015) Encyclical letter LAUDATO SI' of The Holy Father Francis on care for our common home. From https://w2.vatican.va/content/dam/francesco/pdf/encyclicals/documents/papa-francesco_20150524_enciclicalaudato-si_en.pdf. Viewed 9 Apr 2020

J. Singh, *Sustainability and the Sustainable Development Goals in Cultural Clusters*, Sustainable Development Goals Series, https://doi.org/10.1007/978-3-031-70618-9

Gostin L, Koh H, Williams M, Hamburg MA, Benjamin G, Foege W, Davidson P, Bradley E, Barry M, Koplan J, Periago M, El Sadr W, Kurth A, Kavanagh M, Vermund S (2020, August 1) US withdrawal from WHO is unlawful and threatens global and US health and security. Lancet 396(10247):293–295

Hallward-Driemer MT, Hasan AB, Rusu (2013) Women's legal rights over 50 years: Progress, stagnation or regression? World Bank Policy Research Working Paper No. 6616. The World Bank, Washington, DC

Haughton J, Khandker S (2009) The handbook of poverty and inequality. The World Bank, Washington, DC

Hindle T (2009) Triple bottom line–it consists of three Ps: profit, people and planet. The Economist, November 17. From http://www.economist.com/node/14301663. Accessed 12 Feb 2019

Hofstede G (2019) The 6-D model of national culture. From https://geerthofstede.com/culture-geert-hofstede-gert-jan-hofstede/6d-model-of-national-culture/. Accessed 12 Feb 2019

Hofstede G (2020) The 6-D model of national culture. Accessed from https://geerthofstede.com/culture-geert-hofstede-gert-jan-hofstede/6d-model-of-national-culture/

Hofstede Insights (2020) Country comparisons. From https://www.hofstede-insights.com/country-comparison. Accessed 29 July 2023

Kelley C, Mohtadi S, Cane M, Seager R, Kushnir Y (2015, March) Climate change and the recent Syrian drought. Proc Natl Acad Sci 112(11):3241–3246

Kendi I (2020, April 14) Stop blaming black people for dying of the coronavirus. The Atlantic. From https://www.theatlantic.com/ideas/archive/2020/04/race-and-blame/609946/. Accessed 22 Apr 2020

Kerlinger F (1964) Foundations of behavioral research. Holt Rinehart & Winston, New York

Klasen S (2020) From 'MeToo' to BokoHaram: a survey of levels and trends of gender inequality in the world. World Dev 128:Article 104862, 1–10

Langerquest J (2020) Canadian pension funds representing $1.6T AUM call for green COVID-19 recovery. From https://ca.finance.yahoo.com/news/canadian-pension-fund-giants-call-for-green-covid-19-recovery-132358794.html. Accessed 25 Nov 2020

Luhby T (2019) The top 26 billionaires own $1.4 trillion—as much as 3.8 billion other people. CNN, January 21

Mack N, Woodsong S, MacQueen K, Guest G, Namey N (2005) Qualitative research methods overview: a data Collector's field guide. Family Health International, Research Triangle Park

Millennium Ecosystem Assessment (2003) Ecosystems and human well-being: a framework for assessment. Island Press, Washington, DC

Miska C, Szocs I, Schiffinger M (2018) Culture's effects on corporate sustainability practices: a multi-domain and multi-level view. J World Bus 53:263–279

Naipaul VS (1979) India: a wounded civilization. Penguin Books, New York

New York Times (2019, August 6) A quarter of humanity faces looming water crises. From https://www.nytimes.com/interactive/2019/08/06/climate/world-water-stress.html?auth=linked-google. Accessed 4 Apr 2020

Our World in Data (2023a) SDG tracker: measuring progress towards the Sustainable Development Goals. Published online at OurWorldInData.org. From https://ourworldindata.org/sdgs. Accessed 29 Aug 2023

Our world in Data (2023b) Where does our electricity come from. From https://ourworldindata.org/electricity-mix. Accessed 26 July 2023

Oxfam International (2023, January 16) Richest 1% bag nearly twice as much wealth as the rest of the world put together over the past two years. From https://www.oxfam.org/en/press-releases/richest-1-bag-nearly-twice-much-wealth-rest-world-put-together-over-past-two-years. Accessed 27 Aug 2023

Pazienza M, de Jong M, Schoenmaker D (2022) Clarifying the concept of corporate sustainability and providing convergence for its definition. Sustainability 14:7838. https://doi.org/10.3390/su14137838

Puglisi A, Buitendag J (2020) The religious vision of nature in the light of Laudato Si': an interreligious reading between Islam and Christianity. HTS Teol Stud/Theol Stud 76(1):a6063. https://doi.org/10.4102/hts.v76i1

Quinney M (2020) COVID-19 and nature are linked. So should be the recovery. World Economic Forum. From https://www.weforum.org/agenda/2020/04/covid-19-nature-deforestation-recovery/. Accessed 22 Apr 2020

Riessman CK (2008) Narrative methods for the human sciences. Sage, London

Robinson OC (2014) Sampling in interview-based qualitative research: a theoretical and practical guide. Qual Res Psychol 11(1):25–41

Sachs JD, Lafortune G, Fuller G, Drumm E (2023) Implementing the SDG stimulus. Sustainable Development report 2023. SDSN/Dublin University Press, Paris/Dublin. https://doi.org/10.25546/102924

Sharma E, Tewari R (2017) Teaching sustainability through traditional wisdom: a conceptual framework for business studies. Curr Sci 113(2):236–244. http://www.jstor.org/stable/26293855

Singh J (1984) Education and development: operative and espoused views of the school's role in Guyana. J Can Int Educ 13(2):9–21

Singh J (2021) Sustainability: the influence of leadership and culture. Friesen Press, Altona

Solt F (2019) Measuring income inequality across countries and over time: the standardized world income inequality database. SWIID Version 8.1, May 2019. From https://osf.io/preprints/socarxiv/mwnje/. Accessed 12 June 2019

Sridhar K, Jones G (2013) The three fundamental criticisms of the triple bottom line approach: an empirical study to link sustainability reports in companies based in the Asia-Pacific region and TBL shortcomings. Asian J Bus Ethics 2(1):91–111

The Amazon is Approaching an Irreversible Tipping Point (2019, August 3). The Economist, pp 14–16

UN (2019) Sustainability Development Goals. From https://www.un.org/sustainabledevelopment/. Accessed 11 Feb 2019

UN (2020) The Sustainable Development Goals report. From https://unstats.un.org/SDG/report/2020. Accessed 18 Nov 2020

UN (2023) Sustainability Development Goals. From https://www.un.org/sustainabledevelopment/sustainable-development-goals/. Accessed 1 Sept 2023

UN News (2020) UN report sends 'sobering message' of deeply entrenched hunger globally. From https://news.un.org/en/story/2020/07/1068261. Accessed 19 Nov 2020

UNDP (2020) Sustainability Development Goals. From https://www.undp.org/content/undp/en/home/sustainable-development-goals. Accessed 3 Oct 2020

UNEP (2019) Global environment outlook-GEO-6: healthy planet, healthy people. Cambridge University Press, Cambridge

Winfield N (2020, April 19) Pope dreams of post-virus world where inequalities abolished. ABC News. From https://abcnews.go.com/Business/wireStory/pope-dreams-post-virus-world-inequalities-abolished-70230300. Accessed 22 Apr 2020

World Bank (2020) World Development Indicators database. From https://databank.worldbank.org/views/reports/reportwidget.aspx?Report_Name=CountryProfile&Id=b450fd57&tbar=y&dd=y&inf=n&zm=n&country=CAN. Accessed 10 July 2020

World Bank (2022) World Development Indicators database. World Bank, Washington, DC. From http://data.worldbank.org. Accessed 17 July 2023

World Bank (2023a) Country profile Guyana. From https://databank.worldbank.org/views/reports/reportwidget.aspx?Report_Name=CountryProfile&Id=b450fd57&tbar=y&dd=y&inf=n&zm=n&country=GUY. Accessed 30 July 2023

World Bank (2023b) Country profile United States. From https://databank.worldbank.org/views/reports/reportwidget.aspx?Report_Name=CountryProfile&Id=b450fd57&tbar=y&dd=y&inf=n&zm=n&country=USA. Accessed 31 July 2023

World Bank (2023c) Country profile Canada. From https://databank.worldbank.org/views/reports/reportwidget.aspx?Report_Name=CountryProfile&Id=b450fd57&tbar=y&dd=y&inf=n&zm=n&country=CAN. Accessed 31 July 2023

World Bank (2023d) Country profile India. From https://databank.worldbank.org/views/reports/reportwidget.aspx?Report_Name=CountryProfile&Id=b450fd57&tbar=y&dd=y&inf=n&zm=n&country=IND. Accessed 31 July 2023

World Bank (2023e) Country profile Sri Lanka. From https://databank.worldbank.org/views/reports/reportwidget.aspx?Report_Name=CountryProfile&Id=b450fd57&tbar=y&dd=y&inf=n&zm=n&country=LKA. Accessed 31 July 2023

World Inequality Database (2017, 2023) From https://wid.world/world/#sptinc_p90p100_z/WO/last/eu/k/p/yearly/s/false/51.627/60/curve/false/country. Accessed 20 July 2023

WWF (2020) Water scarcity. From https://www.worldwildlife.org/threats/water-scarcity. Accessed 4 Apr 2020

Yeung J (2020, January 13) Australia's deadly wildfires are showing no signs of stopping. Here's what you need to know. CNN. From https://www.cnn.com/2020/01/01/australia/australia-fires-explainer-intl-hnk-scli/index.html. Accessed 1 Apr 2021

Index

J. Singh, *Regional Sustainability and the Sustainable Development Goals*, Sustainable Development Goals Series, https://doi.org/10.1007/978-3-031-70618-9

The manufacturer's authorised representative in the EU is Springer Nature Customer Service Centre GmbH, Europaplatz 3, 69115 Heidelberg, Germany. If you have any concerns regarding our products, please contact ProductSafety@springernature.com

Printed and bound by CPI Group (UK) Ltd, Croydon, CR0 4YY
01/10/2026
02225982-0020